The Garden
at the End of Time

OTHER BOOKS FROM BRIGHT LEAF

THE
GARDEN
AT THE
END OF
TIME

GETTING BY
IN THE AGE OF
CLIMATE CHANGE

John Hanson Mitchell

BRIGHT LEAF
An imprint of University of Massachusetts Press
Amherst and Boston

The Garden at the End of Time has been supported by
the Regional Books Fund, established by donors in 2019 to support the
University of Massachusetts Press's Bright Leaf imprint.

Bright Leaf, an imprint of the University of Massachusetts Press, publishes
accessible and entertaining books about New England. Highlighting the history,
culture, diversity, and environment of the region, Bright Leaf offers readers
the tools and inspiration to explore its landmarks and traditions, famous
personalities, and distinctive flora and fauna.

ISBN 978-1-62534-870-8 (paper); 871-5 (hardcover)

Designed by Deste Relyea
Set in Adobe Jenson Pro
Printed and bound by Books International, Inc.

Cover design by adam b. bohannon
Cover photo by Debbie Costine,
Entrance to the Tea House Allée. Courtesy of the author.

Library of Congress Cataloging-in-Publication Data
A catalog record for this book is available from the Library of Congress.

British Library Cataloguing-in-Publication Data
A catalog record for this book is available from the British Library.

For Jill Brown
The fairest rose in the garden

Contents

Extracts and Evidence

For six days and six nights the winds blew, torrent and
tempest and flood overwhelmed the world, tempest and flood
raged together like warring hosts.
—*Epic of Gilgamesh*, c. 1200 BCE

In the six hundredth year of Noah's life, on the seventeenth
day of the second month—on that day all the springs of the
great deep burst forth, and the floodgates of the heavens were
opened. And rain fell on the earth forty days and forty nights.
—Genesis 6:9

Wasn't that a mighty storm, blow all the people away.
— "The Galveston Flood," American folksong

Ain' gonna be no flood; be the fire next time.
—Nineteenth-century African American gospel song

This is the way the world ends
Not with a bang, but a whimper.
—T. S. Eliot

Sic transit gloria mundi. (Thus passes the glory of the world.)
—Latin proverb

Lord, what fools these mortals be.
—William Shakespeare

Il faut cultiver notre jardin. (We must cultivate our garden.)
—Voltaire

Preface

The Point of No Return

Early in 2022, the United Nations Intergovernmental Panel on Climate Change (IPCC) released a study prepared by more than 270 researchers from sixty-seven nations who were considering the impact of climate change on the Earth in the next few decades. The study concluded that we have reached the point of no return; that the nations of the world are not doing nearly enough to protect cities, coastlines, and farms; and that the dangers are mounting so rapidly that they may soon overwhelm the ability of nature or human ingenuity to adapt. Studies from the World Meteorological Organization have determined that dangerous sea-level rise is guaranteed in the next few centuries, even if the world stops emitting greenhouse gases immediately. A variety of governmental agencies, including the IPCC, concluded that 2022 was the worst year on record for climate-related disasters and was a taste of things to come. According to another UN report, unless drastic international agreements are activated, the current extinction rate will be equal to the devastation caused by the asteroid strike that killed off the dinosaurs some 65 million years ago.

No need to expand on the details of these findings. They're in the news every week: floods, wildfires, droughts, extreme heat waves, rising sea levels, environmental refugees, economic disruptions, and historic extinctions. The point is that we are facing, at an extremely accelerated pace, a new geological and biological age.

So what are we to do? How are we to continue to live contented lives in the face of it all? That cynical old French philosopher, Voltaire, had a solution—just cultivate your garden.

For those who have never read Voltaire's novel *Candide*, or have forgotten the story, the hero, Candide, is an innocent, much-abused child from a minor noble Westphalian family who is caught one afternoon in the arms of the lord's daughter, Cunégonde. As a result, he is evicted from his home and nation. He and his foolish old teacher, Dr. Pangloss, are cast out into a world that, like ours, is falling apart. Candide spends the rest of his days searching for the fair Cunégonde, traveling through war-torn nations led by corrupt governments, filled with bad priests and worse bishops, inquisitors, slaveowners and slave catchers, assassins, and sadistic torturers. He witnesses the 1755 Lisbon earthquake that kills more than 60,000 on All Saints Day morning, when most residents of the city are in churches. Eventually, he does find Cunégonde, but she has suffered a terrible fate: sold into slavery as a prostitute, beaten and broken, she is now old and haggard. Everything that can go wrong continues to go wrong until finally Candide, Cunégonde, and Dr. Pangloss end up on a small farm in Turkey. Of course there are wars all around them, as well as executions, autos da fé, and political assassinations, but the farmer is content to stay home and tend his gardens. Candide decides that this is the best way to survive the absurd chaos of the world—leading to one of Voltaire's most famous aphorisms. "We must cultivate our garden," Candide says to Dr. Pangloss.

Having spent more than fifty years trying to do something about the ongoing destruction of nature, apparently to no avail, I took Voltaire at his word and planted a garden.

This book is the story of the making and maintenance of that garden in the face of climate change. But it is not in any way an argument for inaction or surrender to the disastrous future predicted by climatologists. It is a statement in opposition to the system that created our current situation, a metaphorical act of resistance. The story is a means of gaining perspective on the realities of climate change and also offers a source of refuge and sanctuary, a cure for the increasing phenomenon of climate anxiety. This refuge can be anywhere—from a local woodland trail, to a community or private garden, to even a pleasing vista of green space.

Such sanctuaries are also, in their small way, more than a metaphor. They are a literal counter to the increase in carbon dioxide. Trees are our greatest allies in the control of global warming, as are the herbaceous plants in my one-and-a-half-acre garden, which consists of more than forty species of trees, twenty-seven species of shrubs, and a host of perennial and annual flowers, herbs, ferns, and mosses, all of them madly absorbing carbon dioxide.

The Garden
at the End of Time

PART ONE

Candide's Garden

CHAPTER ONE

THE PEACE ROSE

It's 2045, and here is where we stand:

The winged lions of the Piazza San Marco in Venice look down upon a shallow lagoon. The domes of the cathedral are islands; the bells of the campanile have fallen silent.

New York City is the new Venice. The streets are canals; the buildings are canyon walls circled by crying gulls.

In Miami, vast king tides sweep over the neighborhoods of the city twice a day. The Florida Keys have disappeared.

Galveston has flooded.

New Orleans is a bayou plagued with alligators.

The Arctic Ocean is almost ice-free. The glaciers of the world are half gone.

Low-lying islands of the western Pacific have disappeared.

Bangladesh has disappeared; half of Pakistan is under water.

The forests of North America are thin and scorched and ravaged by fires and insects.

The American prairies are deserts.

And I am but an old man in a dry month, paying out the end of my days and looking back on that singular year when it all began.

It was the spring of 2022, and out in the garden old goat-footed Pan jumped down from his pedestal in the boxwood close and danced off through the garden paths, piping all the way, along with the dawn chorus of the birds.

The local woods were breaking out in green that day, and everything was crowded together—the flowerbeds and the early roses, the dogwoods and the yellow pennants of the laburnums, the cherries and the crabapples, the Carolina snowbells, the lilacs, the rhododendrons, the azaleas and viburnums, all of them blending to create a wild abstracted palette against the various shades of green in the leafed-out trees and the lawns.

All the world (my world, at least, not the real world beyond the garden walls) was singing: the robins, the towhees, and the mockingbirds joining the chickadees and house finches, the pine warblers and titmice, the cardinals and Carolina wrens. And down in the woods where the old white oak stood, gray tree frogs joined the chorus, and I could smell the essence of life rising out of the moist earth and the fresh-mown grass and the scent of roses, the scent of lilac, of lavender and crushed thyme.

And I lived there too in that garden, successfully disguised to myself as Pan, or the Green Man, who dressed in ivy and lived in the forests of medieval Europe and deigned to come into the gardens from time to time. Bliss it was that spring if I stayed within the green walls of the garden. But out beyond the garden that year things were not going well.

There were uncommon floods throughout Europe in the early spring. One whole village in Germany was swept away— rooftops under water, people drowned. Horses, cows, pigs, and goats disappeared in the swirling brown waters, their bodies strewn over the riverbanks.

Here in North America, there were major prolonged droughts in the Midwest, followed by firestorms, floods, and mudslides. The Covid pandemic was lurking in the shadows. People were dying. In the United States there were periodic massacres of children. The concept of democracy was threatened, and all around the world repressive governments were dismantling environmental laws. Refugees were everywhere, fleeing the floods and the droughts and the governments that were intent on killing people of different religions or skin color or political persuasions.

There were record heat waves in China, Europe, and the North American West; and in Pakistan 30 million people were uprooted by a mega monsoon that flooded one-third of the country. It seemed as if the very Earth was either burning or drowning.

Even here in New England, which was said to be one of the least affected regions of North America, the three rivers near my garden flooded their banks in the early spring. This was nothing compared to the deluges out west, but the rising waters spread over lawns, filling cellars and ruining garden plots and newly planted fields.

The first heavy rains had come in with high winds and lightning strikes, knocking down trees, smashing houses and garages. There were widespread power outages, and roads were closed due to downed powerlines. Even after the storms passed, the three local rivers continued to rise. Then, following the flood, came a record drought.

What to do?

I went out and bought another rosebush.

This had become my habit in recent years. A disastrous wildfire in California? Buy a rosebush. A record-breaking drought in the Midwest? Get a rose. Devastating, climate-related floods in Bangladesh? More roses. Record-breaking heat waves worldwide? Plant more roses.

There were so many climate disasters in 2022 that I began to run out of space and had to dig out new garden beds. And it was in this manner that the garden began to expand, in spite of the fact that I barely had time to take care of the already established grounds, which by that time were more than thirty years old and covered nearly an acre and a half—a world of ornamental trees and shrubs that required constant attention in the growing season.

I found a healthy Peace rose in a nearby nursery that has a very good selection of rosebushes. Fittingly, this particular rose was developed in 1935 just before another global disaster—the Second World War. It was cultivated by the French rose breeder Francis Meilland, who saw it as a promising and interesting new variety. But 1935 France was not the safest place to go about the business of cultivating a rose variety, a process that takes years of growing seasons. So Meilland shipped several slips of budwood to growers outside of France. True to his fears, Hitler invaded France, and the world was turned upside down. But a few of the budwood slips had been smuggled to America, where a rose grower planted them and gave them to other growers to test. The new variety did well, and other propagators spread the slips throughout whatever agricultural zones they could survive in. In April 1945, a month after Germany surrendered, the new rose variety was officially named the Peace rose.

The Meilland family had no idea what had happened to their slips until after the war.

I took my new rose home and planted it in a section of the garden where I used to grow herbs until I started buying roses after every major worldwide disaster and needed the space. My project didn't slow the great shuffling beast of climate change, nor the mass murders, nor the election of dangerous national and international leaders. But it was helping to blur the reality.

The rose bloomed two weeks after I planted it and, to better appreciate it, I pulled a chair up and sat in the sun, watching it, as if somewhere in the varying shades of yellow folds I might find a key to world events.

A bee landed and began buzzing around the pistils of the flower, sipping nectar and gathering pollen. A Carolina wren sounded off, announcing something, or seeking some beautiful Jenny wren perhaps.

This was late morning, late May, clear and warm, and I fell into a state of half sleep in which I could hear other bees working the flowers nearby. Sparrows were chirping and flitting in the shrubbery.

A blue jay called.

I could smell moist earth, grass, a faint scent of flowers. A whisper of a breeze set the poplar leaves fluttering, and a spring azure butterfly drifted past, a sign that it would soon be summer, with a rising heat, shimmering fields, rain, more rose blooms— summer afternoons in a green shade.

I slipped into one of those comfortable states of displacement I occasionally experience in this garden and found myself somewhere in my past, in a warm spot in an unidentifiable country. And then that scene dissolved, and it was a summer afternoon in England, and there was a river below the meadows, and there is nothing quite so fine, in a civilized sort of way, as the taking of tea on a summer afternoon, so I was taking tea in a daydream.

There was a silence in the home garden, save for the bees in the roses.

The peace of wild things.

And then the world exploded.

Out on the road below the house a huge dump truck hit a pothole with a metallic crash, shattering all daydreams and spewing carbon-loaded exhaust into the bright blue skies of May.

A taste of reality.

Only fair, I suppose. One should not be so comfortable in our time.

CHAPTER TWO

THE RAMBLER ROSE

I had come to this property thirty years earlier, after poking around in gardens and estate grounds in Europe. At one point—why, I don't remember—I found myself on the island of Sao Miguel in the Azores. Here, in the little inland town of Furnas, I happened upon an old formal hotel with extensive gardens that had once served as the nursery for Kew Gardens, in London. The grounds and gardens, which were well tended but no longer serving as a nursery, surrounded a hot spring where a troupe of northern Europeans had gathered, taking the waters. It was a quiet spot, set apart from the noise of the industrialized world (quite far apart since the Azores lie in the mid-Atlantic).

Elsewhere in the archipelago, hydrangeas had established themselves. They had grown so lush and so exuberant in the favorable climate that they swept the windows of the local buses traveling along the winding inland roads. On the island of Terceira, I saw little plots of household gardens growing beds of immense, large-leafed kale, and at one point I came upon an unlikely walled garden where, of all things, artichokes were growing.

Ever curious, I climbed over the wall to have a better look, and here I met a little man dressed in somewhat soil-stained trousers, a vest, a collared shirt, and a tie. He was bent over a hoe, tending the plants.

In my broken Portuguese, I greeted him and we fell into conversation, and after a while he invited me to meet the owner, who, he assured me, was a very pleasant woman who enjoyed visitors. (At least I think that is what he said. It was possible that I was about to be reprimanded for trespassing.) In any case, I followed him up through the artichoke beds and treed allées to a terraced villa with tall French doors that faced florid gardens of rhododendrons and hydrangeas. Muddy boots and all, he showed me to a cool shaded library lined with shelves holding ancient books, along with a lot of antique brass pots and houseplants, and here I was introduced to a handsome woman with copper-colored hair and Gallic features. She greeted me in softly accented English and invited me for tea.

"We do not see many visitors here," she explained. "Especially not people who have any interest whatsoever in the gardening."

Tea served, we chatted about island life, the local *vinho verde*, Atlantic seabirds, the gardens at Furnas, and also—and at length—her family.

She was a descendant, she said, of early European settlers of the Azores who had come to Terceira in the early fifteenth century from Bruges, in Flanders, which would explain her copper-colored hair—not a common hair color on the islands, which were settled mainly by dark-eyed Portuguese and Spanish colonists.

Later, I traveled through the archipelago on the little inter-island steamer, stopping off at all of the islands, including the most distant and isolated, Corvo, which is the farthest west of the nine islands of the archipelago. I came across more gardens and vineyards during the passage, and I met, as one often does on isolated islands, a number of unique characters, including an off-center gentleman, the self-appointed King of the World, who was then living in exile on Corvo.

I seem to have had in those years an affinity for islands. For the better part of a year, I worked in a restaurant and pension on Corsica, which ironically, given my later interest in roses, was named the Rose Café. Before that, I spent a few summers working on islands off the New England coast, and on my return from the Azores I took a house-sitting job on Martha's Vineyard caretaking a large estate garden that would later become the Polly Hill Arboretum.

I'm not sure whether it was that work or the influence of the Azorean gardens or an inspiring visit to an isolated monastery garden in a remote hill village above Sorrento that encouraged me to create an ornamental garden of my own.

All I needed at that point was land. That settled, I eventually did manage to get a property in the middle of a group of five working farms located thirty-five miles northwest of Boston on grounds formerly controlled by a powerful Native American chieftain named Tahattawan.

This unique square-mile tract had a long history of farming that stretched all the way back to the 1650s—far longer if you count the 7,000-year history of the Native people who grew their crops there. In the nineteenth century, for some obscure reason, the locals termed this section of the town Scratch Flat, and here I finally settled in and began to plant.

The section of Scratch Flat where I lived was a former apple orchard; in fact, in the early nineteenth century there were more apple orchards in the town than in any other town in the state. But save for a grove of hickory trees interspersed with a few ancient crippled apple trees, the plot I owned was covered with seventy-year-old white pines. So the first thing I did was to cut down the pines—what might be seen as anathema to a supposed environmentalist such as myself. But thick stands of a white pine create what ecologists call depauperate land. The forest floor of fallen needles, coupled with the deep shade created by the overstory, not to mention the acidic soil, does not create a favorable environment for a diverse understory of deciduous shrubs and flowering herbaceous plants.

Before I decided to cut down the trees, I did a survey of the plot and, except for beetles and soil-dwelling invertebrates, counted only five species of higher plants, including large beds of poison ivy along the sunny edges. No birds nested, and I found only one squirrel nest there. I came back in winter and reviewed the tracks of mammals and found evidence of coyotes, fishers, possums, raccoons, and foxes but no dens or nests; the animals were just passing through on their way to better hunting and foraging grounds.

At the time, I happened to know a crew of foresters who used mules rather than tractors and bulldozers to haul out the cut trees, so I hired them.

No heavy carbon-spewing machines.

Once the land was cleared, I went to work digging new beds, creating over the years garden rooms and winding paths and plantings of perennials and flowering trees and shrubs in and around the extant deciduous trees.

I started with a small monk's garden of herbs, a few vegetables, and a bed of annuals and then got interested in delphiniums, then began collecting daylilies from a neighbor, and then finally began shaping the grounds into a semblance—a very small semblance indeed—of a somewhat disheveled Italian Renaissance garden, complete with allées, garden rooms, a labyrinth, meandering paths, and finally, in its latest iteration, beds of roses.

I did hire a friend with a rototiller to prepare some of the beds, but other than that I carried out most of this work by hand.

As far as I'm concerned, there was good reason for that. Physical work without the use of internal-combustion machines is just one positive, earth-friendly aspect of the art of gardening. For one thing, working with hand tools slows you down. Time and again, after a day's work in the garden, when I would venture out into the real world beyond the gardens, everything seemed to be moving at a high rate of speed, mainly the cars, but also people rushing here and there as if late for some event.

But there is a larger almost spiritual feature of gardening by hand. Manual labor puts you into close contact with earth, air, sky, and water. Peer-reviewed studies have determined that mere exposure to green space, even without any activity—just sitting on a park bench looking at trees, for instance—is salutary. Sunlight is beneficial; it is the best source of vitamin D. A mere twenty minutes in the sun helps regulate the melatonin levels that contribute to sleep. Longer exposure causes a rise in serotonin, which is calming and can evoke a positive mood. And according to a recent systematic review by the Japanese research group Environmental Health and Preventive Medicine, forest bathing, or Shinrin-Yoku—a slow walk in a forest—induces a calming effect and helps lower anxiety by reducing stress-related cortisol levels as a result of exposure to the volatile organic compounds known as phytoncides, which are released by shrubs, trees, and garden plants. And finally there is one of the greatest health benefits of all: what one member of my circle, a doctor, calls "the elixir of youth"—exercise.

Much of the daily work in the garden is a series of small jobs, such as pruning and planting and deadheading flowers and the like. But of all my various chores the most satisfying is to cut the rough overgrowth of clovers, ajugas, and so-called weeds in the meadow in front of the house with a scythe. This is hard, but it is slow rhythmic work in which it is easy to drift into an almost meditative state.

I first learned the art in the Azores, where I watched a groundskeeper for a government building cut a smooth green lawn with a sharp scythe. He moved slowly across the lawn, keeping the tool's heel close to the ground and sweeping the grass seemingly without much effort up toward the building, then turning and cutting down toward the road, where I was sitting on a park bench watching him work.

I noticed that he would stop periodically after a few passes, review his work (I might say proudly, but that would probably be an exaggeration), and then slick the blade with a sharpening stone he carried in a pouch attached to his belt and carry on.

When he finished his work, he stopped not far from me, and I took the opportunity to talk to him. I had tried scything earlier on a friend's farm in Maine, cutting down a weed patch, but after one pass I was tired out. I explained this to him and asked how he managed to keep going so effortlessly.

"Work slowly," I think he said (this interview was in broken Portuguese), "and keep the blade sharp and run the blade low across the ground."

I took his advice when I got home but couldn't quite get it. I still don't, but I did learn that if I worked slowly and stopped often, I could at least keep working. And stopping often turned out to be a good way to make deeper contact with where I was standing, to take in the moment: the smell of fresh-cut vegetation; the song of the indigo buntings, which seem to be active in mid-June; and also the time and place in which I was currently fixed. It offered a good opportunity to place myself in the context of history.

The scythe was invented, in one form or another, in the twelfth century; and until the advance of mechanized agriculture in the mid-nineteenth century, it was the tool of choice for reapers.

One year, when I was cutting the meadow in June, this continuity of union with the past struck me with particular force. Standing in the surround of Quaker ladies, violets, ajugas, and clovers, resting between sweeps, for some reason I thought back to Tolstoy's description of scything in *Anna Karenina*. His protagonist, Levin, is out scything the wheat fields, side by side with his peasants. After struggling and tiring himself out with the work, he begins to observe his fellow workers, who seem to be able to cut steadily for hours, with only a few rest stops. After a while, as he works at their pace, he slips into a meditative state of suspended time in which he finds himself in union with the scythe, the wheat fields, and his fellow workers—lost in the moment.

As far as exercise is concerned, a few years back I measured the distance one could walk within this acre-and-a-half plot of land. Starting at the head of the long driveway, I circulated around the outer walled edges of the property and then began following the paths inward, weaving here and there, coiling through the courses of a labyrinth I had created and emerging into the various garden beds, walking ever onward and inward until I reached the house.

All told, if you follow each of these circling paths, you end up walking for more than a mile—much more if, as I do almost every day, you spend a couple of hours walking around the garden doing all of the necessary chores of maintaining your plantings and then ranging out over other open spaces surrounding the property.

There is another generally unappreciated or unrecognized benefit to creating a garden, especially in our time. Planting is a statement of belief in a future. Growing vegetables is a seasonal commitment, a cycle. But laying out a large ornamental garden or a landscape with trees and shrubs is a long-term connection with a place. It is an engagement with what the Spanish call *querencia*, a personal involvement with the seasons and the natural history of your territory. It is also a statement of what the Hopi call *tuwanasaapi*, or "place where you belong."

A garden anchors you in the land, and that means you will establish a connection with your location and your region. It means you will be more likely to stay put and work to save what you have rather than pull up stakes and light out for the territories when things get bad.

Finally, a garden is literally a biological act of resistance to climate change. As a result of my radical act of clearing a forest of white pines and replanting it with a garden, this one-and-a-half acre tract features more than three hundred trees consisting of more than forty species as well as fifty or more shrubs, including twenty-seven different species, plus a host of perennial and annual flowers, mosses, herbs, ferns, and grasses, with all of these plants absorbing carbon dioxide every day and creating a biodiverse, albeit small, plot of land.

After some twenty years, when the garden came into its own, I did another ecological survey. I counted all of the different species of trees, shrubs, and perennial plants I had grown and also the numbers of birds, mammals, reptiles, amphibians, insects, arachnids, fungi, lichens, and mosses. Altogether there were more than 2,000 living things on what had been a dry stand of white pines with only five or six different flowering plants.

In connection with all of this, my choice to plant roses to memorialize disasters was not arbitrary. The rose is one of the most heavily laden symbolic plants in the world: beautiful, richly scented, yet barbed with thorny stems. The flower is a story in itself of beauty and danger, love and death, sex and virginal purity.

Roses feed your soul. They have been feeding humanity's soul for thousands of years. Wild roses, which grow virtually everywhere in the temperate climates, were first cultivated at some lost point in history after the development of agriculture, but they were undoubtedly appreciated earlier or no one would have bothered to tame them. This singular genus—*Rosa*—counts itself among the elemental symbolic forces of nature along with the sea, sky, sun, stars, moon, birds, and the lion.

Roses were probably first cultivated in about 750 BCE in the time of the Median and Persian cultures. These early cultivars were only a slight improvement over the wild rose, plants with a single-petal flower that probably produced more blooms than its progenitor. Their earliest extant mention is implied in Homer's repeated epithet describing sunrise as "the rosy-fingered dawn" in the *Iliad*. Roses appear again in the third century BCE in one of the Greek poet Sappho's poems. She characterizes the rose as "the Queen of Flowers," an epithet that reappears throughout history and is still used in our time. Roses are mentioned in the King James translation of the Bible, in Isaiah 35:1 (although the famous rose of Sharon is not a rose but a member of the mallow family).

But it was the Romans who first began serious cultivation of the plant. I can't remember seeing roses in the highly realistic mural in the remains of Queen Livia's villa outside of Rome, but they must have been depicted there. Practically every other wild and cultivated plant appears in that epic mural, which covers three walls of the queen's dining room.

Virgil mentions roses in book 4 of his long agriculture-based poem the *Georgics*, and they figure in the myth of the death of Adonis, who was gored by a boar: roses grew in the spots where he spilled his blood. The Romans even had a May celebration called Rosalia, which celebrated the rose. It was held to honor the dead but was also symbolic of rebirth, rejuvenation, and memory. And interestingly, as far as this book is concerned, roses also grew in Emperor Julian's garden near Constantinople. Voltaire knew of this garden and used it as a model for the farm where Candide determines to stay put.

By the Middle Ages, the rose came into full literary bloom (so to speak). The crusaders introduced the plant to Europe, and it was during this period that the flower took on Christian symbolism, with its five leaves standing for the five wounds of Christ and the white bloom symbolizing Christian purity. The flower is mentioned often in the art and literature of the period and was used as a symbol in the lays of courtly love and other poems and songs of the Provençal troubadours.

One of the most important and popular works of the period is a long allegorical poem called *The Romance of the Rose* in which, in a dream sequence, the main character, known as the Lover, seeks union with a beautiful chaste maiden. The poem deals with all of the conflicting themes of medieval literature: love, reason, sex, and Christian purity.

In *Paradise*, the final book of *The Divine Comedy*, Dante also relies on the symbol of the rose, which he associates with the Virgin Mary, grace, and divine love. By contrast, in the profane works of the popular eleventh-century lays of courtly love of the troubadour poets and musicians, the flower is identified with the conflict between spiritual attainment and erotic desire, passion, beauty, and sexuality all mixed together. By the fifteenth century, the rose had been politicized, notably in the Wars of the Roses. In 1688, the Jacobites continued this political tradition, choosing the white rose as their emblem. The rose bloom was still a plain single-petal flower, although by this time it was available in a variety of colors, as the white rose of York versus the red rose of Lancaster makes clear.

By Shakespeare's time, cultivators had developed the musk rose, a many-petaled bloom with a musky scent. But the biggest change in cultivars came after western nations opened trade with China in the late nineteenth century. Roses had been cultivated in China for centuries, and plant collectors were now able to bring new varieties to Europe and the Americas, including the Father Hugo rose (still popular today) and the Golden Rose of China, a hardy early-blooming shrub rose. Since that time, through cross-pollination and selective breeding, cultivators have bred more full-flowered plants with extended flowering seasons, and by the twentieth century they developed the constant-blooming exhibition flowers we know today.

For my part, all this messing about in the garden while the Earth burns is not the idle hobby of a landed country gentleman with enough money (which I do not have) to waste on roses and new trees. But the fact is I see gardening as an act of defiance. As the American ecologist Aldo Leopold argued, a thing is right when it works to preserve the stability and integrity of nature; it is wrong when it does otherwise. So gardening might be said to be an example of the Buddhist concept of right action. It is a resistance to greed and the capitalistic mad-dash world of getting and spending. It is also—and this is my main point—a means of finding peace of mind, which has been a tradition in both eastern and western cultures for centuries—for instance, with the aforementioned Japanese tradition of forest bathing.

The Roman poet Horace fled the politics of Rome and retreated to his garden. The Chinese scroll painters and poets of the ninth century deserted civil life and fled to the rounded karst mountains of Guilin to lead quiet lives close to nature.

Voltaire fled to Ferney, outside of Geneva, and planted a garden.

Vincent van Gogh retreated to a garden to escape whatever demons were haunting him. In 1869, he committed himself to the Saint-Paul de Mausole psychiatric institution in Saint-Rémy-de-Provence. Here, among its patterned gardens and allées, he found solace and comfort and was also inspired to create what some critics consider to be his best paintings.

The Norwegian artist Edvard Munch, troubled by the early deaths of his mother and sister, is best known for his morbid depictions of illness and loneliness. But later in his life he retreated to nature to soothe his anxieties and turned to a spiritual relationship with forests and the sea. In his last years he purchased an estate on rolling lands with apple orchards, planted a nursery, and lived there until his death in 1944, almost happy.

Henry David Thoreau went to the woods to live deliberately and there grew green beans in a clearing beside Walden Pond. Thoreau wrote that he planted beans but harvested metaphors. By planting a garden, he claimed that he had established a balance between wild nature and the civilized world of nature restructured. By hoeing and harvesting, he fixed himself in two worlds—the wild untamed wilderness he so loved and praised and the organized constructed world of agricultural civilizations. He wrote that, in order to preserve his health and spirits, he felt it necessary to spend at least four hours a day sauntering through woods and fields to free himself from worldly engagements.

I suppose in some ways I was replicating Thoreau with my own garden—planting roses and harvesting metaphors.

In Thoreau's case, as in so many others involving gardens, including van Gogh and Munch, this contact with nature was also a balm for personal disaster. Henry's brother John, to whom he was close, died of lockjaw the year before Henry moved to Walden. His two-year experiment in the woods offered a solace that was unavailable in his normal daily town life.

Thoreau was not the only writer who saw metaphors in growing things. One afternoon in the spring of 1939, while Virginia Woolf and her husband Leonard were living at Monk's House in Bidwell, Leonard was out in his orchard planting irises when he heard Virginia calling from the sitting-room window to tell him that Hitler was giving a speech on the radio.

"I shan't come in," Leonard shouted. He said the irises he was planting would be flowering long after Hitler was dead. And in fact, twenty-one years after Hitler committed suicide, they were still blooming each spring.

Literary and diplomatic careers notwithstanding, Leonard Woolf was a serious gardener; and when he finally got some land of his own, he went at it obsessively. "I'm always losing him in the garden," Virginia wrote to a friend. (My own wife says this about me, incidentally.) Virginia herself experienced two brief periods of happiness in her troubled life—first in the flower gardens of her youth, then again in the gardens at Monk's House.

After Virginia's suicide, when, like Thoreau, he needed refuge and solace, Leonard carried on, clipping his hedges, tending his fruit trees, and cultivating the six acres of land at Monk's House. He gardened to the end of his days and listed gardening as one of his life's pleasures in his autobiography.

Ironically, the social critic and novelist George Orwell said more or less the same thing. He led the kind of dramatic life one wouldn't necessarily expect from a gardener. Among his many experiences, he joined the Republican Army during the Spanish Civil War, taking a bullet to the throat while fighting on the Argon front. Yet he cited gardening as one of his abiding interests, including the cultivation of roses, which he planted wherever he happened to find himself. Between 1946 and 1949 he lived in Barnhill on the island of Jura in the inner Hebrides, and even in this windy, rocky environment, he managed to nurture plants in his thin hardscrabble yard. He wrote his best-known political novel, *Nineteen Eighty-Four*, while he lived there.

The environmental writer Rebecca Solnit fills out the ironic details of Orwell's gardening experiences in her 2021 book *Orwell's Roses*.

Unlike Virginia and Leonard Woolf, we are not currently at the precipice of a worldwide war (though in 2022 there were rumors of a brewing second American civil war, and there was a nasty war in Ukraine that at the time was threatening to spread). But most adults are spinning out their years in the face of a catastrophe of a different sort, one that is not obvious to most of us in our daily lives—until the river rises or the house burns down. We are now living in a period of climatic history, in which ruin seems imminent. If there is a literate future (which some cynics doubt), the history books will cite our time as a major geologic turning point in which land forms reshaped themselves, ancient glaciers melted, civilizations collapsed, and massive extinctions similar to those in past epochs altered the natural history of the planet.

We need gardens now more than ever. We need contact with nature more than ever. We need more urban tree-planting programs, more green parks, more community gardens, more vacant urban blocks undergoing renewal and replanting. We need the grand and ambitious plans of so-called ecological artists and land-use planners. We need regenerated landscapes. We need to support projects that will allow access to green space on a regional scale so that, despite the sprawling, overcrowded, ever-increasing refugee-spawned, chaotic urban areas of the world, everyone can find peace in a green shade.

In short, we need the metaphorical garden.

CHAPTER THREE

TO A WILD ROSE

On any given clear morning in the growing season, I tend to follow the sun around the garden. It first lights up flowerbeds at the western corner of the property, where I have a small greenhouse and a fishpond surrounded by young dogwoods, Japanese maples, and a variety of flowering perennials. I have set out chairs and a table in a sunny glade there, and I sit for a while, taking in the warming sun, drinking coffee, and sometimes sketching out new garden designs or writing, but more often just staring into space, thinking of nothing.

As the sun begins to sweep across the main gardens, I move on, pass down a long, tunneled allée of climbing roses to more flowerbeds, past another fishpond, and on to a circle of small flowering trees and then down another path lined with peonies and blue stars to a circular lawn bordered by beds of roses. I carry on, following the sun, passing other flowerbeds and allées and into the courses of a labyrinth I laid out years ago, then back toward the northwest, where I have a little studio based on Thoreau's cabin at Walden.

More wandering as the sun sweeps southward: past flowering trees and shrubs and all through the gardens to my version of a carriage house, where New Dawn roses climb the bay walls. I pass more flowerbeds along the driveway and come to a gated secret garden in front of the main house. Then, in late afternoon, I return to the greenhouse garden to watch the sun set over the farm fields to the west.

This garden lies at the center of a surround of four other local gardens and, on the western slopes of the hill where my house is located, two working farms. In fact, these gardens are encompassed by a sort of Jeffersonian landscape, a square mile of woodlands and working farms that make up Scratch Flat.

To the south, there is a garden laid out and managed by a retired French teacher who, ever since childhood, has had an interest in gardening and garden history. The land was formerly owned by the groundskeeper for one of the old Boston Brahmin estates, and the soil there is particularly rich. My friend André (his real name is Andrew, but that does not reflect his Francophilian leanings) began by establishing a small walled garden near the porch of his tiny house. Then, year by year, he expanded the garden spaces of the two-acre property, adding, among other features, a large garden room defined by walls of forsythia bushes as well as a pseudo Roman temple, an arched wisteria-covered gazebo, and his wide collection of native and non-native trees and shrubs.

North of André's garden, an Italian gentleman maintains a fertile and productive vegetable garden along with fruit trees and a surround of shrubs and flowering vines. Next to him is another productive vegetable garden as well as fruit trees and an abundance of daylilies managed by a woman who, among other things, happens to be my ex-wife. She also owns an acre or so of agricultural land on the other side of the road that she lets out to a young aspiring local farmer.

My own garden is just to the west, set on a rise above that tract of land. Then, to the north, there is the substantial, well-maintained garden and grounds of my daughter and her family.

True to the Renaissance garden model, I have let a section in the north part of my garden go wild. There is a vernal pool in that area and a stand of red maples, slippery elm, and a few black oaks and a hemlock tree. This section gives onto a tract of private property that is essentially the defining garden of this whole neighborhood, a twelve-acre plot of managed forestland, ornamental trees and shrubs, and flowerbeds. The forested section of the land consists mainly of a mixed woodland of red and black oaks, red maples, white pine, and a dark grove of hemlocks interspersed with nearly one hundred varieties of rhododendrons, azaleas, and native shrubs and wildflowers.

Rick Findlay, the owner of this semiwild garden and the theoretical lord of the manor, is a retired landscape designer and a well-known local environmentalist who works diligently with the local land trust to save as much open space as he can. In fact, not unlike Lorenzo di Medici, who sometimes toiled in his garden alongside his peasants, Rick has spread his conservation efforts over a wide range of public lands, many of which he had a hand in preserving. He clears trails, and oversees the construction of footbridges and the like, but his main work in our time is a (probably vain) attempt to clear the land of invasive plants.

His property, which was split up sometime in the 1930s, originally included four of the five gardens in the neighborhood as well as the orchards to the southwest that made up one of the original five farms of Scratch Flat.

The forested section of Lord Findlay's land is laced with trails that wander in and around the darkened tree boles; and if you follow the outermost path, just across the road from his property there is another larger section of forest that runs along the banks of Beaver Brook, a local stream that flows through a wide cattail marsh and gives onto a lake and defines the eastern boundary of Scratch Flat.

This streamside tract of wild land has an entirely different ambience from the other properties. The floor of the forest here is rich with ferns and little rivulets that run down through the land and empty into the marshes of the stream. If you bushwhack off the main trail away from the brook, the makeup of the land changes: deciduous trees open to sunny glades and boggy springs that feed the rivulets on the lower slopes.

When I first settled here, I discovered this wood on one of those spring days when the fresh earth was just coming into flower and the air was filled with the twitterings and songs of Carolina wrens and robins and the buzzing calls of passing warblers. In the clearings, filtered sunshine was blinking in the breeze that fluttered through the upper story of the trees. A green moist light suffused the air, and there was a warm scent of unidentifiable flowering plants. The floor was strewn with celandine poppies, foam flowers, bloodroot, mayflowers, and lilies of the valley, and I wandered on until I came to an area where the woodland flattened out and was shaded with an overstory of maples and oaks.

The whole atmosphere in that grove seemed to be unshackled from time and place, an enchanted forest, decidedly different from the world beyond the glen, where there were roads and encroaching suburbs and a strip of power lines. Here at the edge of the glade I found the source of the sweetened air, an old lilac shrub in full bloom. Nearby I saw a section of daffodils that had just completed their spring flowering and, a little farther on, a bed of ancient peonies nearly overwhelmed by flowering blue periwinkles. It was clear that this place had been inhabited at some point. And poking around the woods, I eventually found the remnants of a house foundation.

I am, among my other eccentricities, an habitué of ruins. In my earlier wanderings in southern France and Italy, in Mexico and through Central America, and even on some of the islands of the South China Sea, I have often accidentally come across unpopulated ruins, many of them well off the beaten tourist tracks of more famous sites. Within the dense covering of trees and vines that usually overwhelm a good ruin, I can read a message of the enduring power of wild nature.

But other than a few purported Native American sites, most of the ruins in the American Northeast consist of random stone walls winding through second-growth woodlands and old house foundations, root cellars, and walled pounds where stray cattle were once held.

This foundation, with its old flower beds overrun by periwinkle, was clearly once an inhabited—perhaps even loved—house and garden, now long deserted. And herein lies another climate-change story, albeit of a different sort.

Until the 1960s, the little town on the east side of the brook was known for its agricultural products, mostly apples but also, in the late nineteenth century, for its beef and wool and its vegetable produce, which supplied Boston's open-air Haymarket. The juice company Veryfine was established there in 1865 because of the plethora of apple orchards.

Farming never was easy in New England, but it got worse after 1815 when the Mount Tambora volcano in Indonesia blew its top. It was the most powerful volcanic eruption in recorded history, and the aftereffects of this mighty blast created devastating climatic conditions. A blanket of ash spread over the world, shading the sun and causing temperatures to decline worldwide. New England had killing frosts in July, then more in late August, creating what became known, in 1816, as the year without a summer.

Unfortunately, the ruins wrought by Mount Tambora were nothing compared with what was to come.

Even before the eruption, New England farmers had been moving westward to more fertile lands in New York State. Then in 1825, engineers opened the Erie Canal, making escape even easier. As far as I can tell, the family that owned this particular

tract of land moved west around 1850. Judging from the archeological evidence I unearthed in the ruin, they apparently gathered
what they could transport, shut the door, and left their house
empty. In time the doors blew off and the windows shattered.
Heavy snows broke the back of the roof; the walls caved in, and
slowly, little by little, the house returned to ground. The same
thing happened to a small home next to my house, a place that
was once owned by one of the Scratch Flat farmers.

But in both cases, the natural world endured. Down in the
old woods, raccoons moved in and barn swallows nested in the
front hallway. The family's garden plots, orchards, and hayfields
grew up in grasses and forbs, which were followed by sun-loving
shrubs and trees such as blueberries, gray dogwood, and white
pine, the tree that came to dominate parts of the landscape in
this region. But in the case of this particular plot—for reasons
probably related to soil conditions—the deciduous forest that
had dominated the land before the trees were cut somehow
returned intact, creating the woodland I discovered when I first
settled in the area.

And all the while, as the old world warmed and floods and
droughts advanced, my little world of gardens and woodlands
became all the more precious, a lone reminder of the endurance
of the ancient forces of nature and an uninterrupted round of
the four seasons.

Plants and animals survived in the little grove that could no
longer be found in other parts of the town. I once discovered the
nest of the secretive blue-gray gnatcatcher there. I have heard the
now-rare hollow fluting calls of the veery and the wood thrush.
Yellow warblers nested near the stream; kinglets and marsh
wrens flourished in the nearby floodplain; and every spring, just
before the leaves of the oaks and hickories and maples broke out,
a host of wood warblers would pass through.

When I first arrived, Canada mayflowers covered the ground in
some areas, and lilies of the valley bloomed, along with bloodroot

and mayapple. I saw Dutchman's britches growing there once
and squirrel corn and hepatica and other wildflowers that I
have never found except in a nearby arboretum that maintains
a native wildflower trail.

Summers brought on a close stillness that was almost tropical
in the sheltered rooms, and at dusk under the shadowed halls,
I could almost imagine myself in some ancient land set apart
from the world beyond. In time, at least for me (and I seem
to have been the only person that regularly visited the little
hollow), the woodland took on an almost mystical presence. I
imagined sometimes the rhythmic chanting and slow drumming
of ancient native shamans, mixed with the dull booming of great
horned owls, the echoing, hollow call of the barred owl, and the
drumbeat of grouse wings.

The place evoked the mythic denizens of the classical dark
forest of the western imagination: witches' covens, fairy rings,
the storied yelping of the hounds of the wild huntsmen sweeping
over the tree tops and carrying off innocent farmers, a cavalry
passing overhead, bridles jangling, swamp demons rising out
of the misty wetlands, trolls and naiads—all things seemed
possible there, especially at night. It was another country, a
place apart.

(To be fair, no one else seems to have experienced such things.)

Needless to say, this was not the first period in history that experienced political upheaval and natural disaster. But when such
events take place, they seem to have the counteractive effect of
encouraging the creation of new gardens. The great gardens of
the Italian Renaissance were established during the mafia-like
rule of the Medici family, and Versailles was laid out under the
decadent but powerful reign of the Sun King, Louis XIV, a time
when Catholics and Protestants were hard at work slaughtering
each other. Some of the eighteenth-century urban walled gardens

of Italy and France came into existence partly because there was so much violence in the streets. Paris, for example, suffered through seventeenth-century versions of road rage, when carriage jams and minor insults led to outright duels in the streets.

One of the model examples of a retreat of this sort was Voltaire's garden, Les Délices, in Ferney in eastern France. Old Monsieur Voltaire, who was in his mid-seventies when he began work on *Candide*, might have appreciated some aspects of our own coming climatic change. He was always cold. His estate was in the Jura, not by any means the warmest department of France. But it was right next to the Swiss border; so when the French government came to arrest him, which they periodically would do because of his radical and anticlerical writings, he could duck into Protestant Switzerland.

He sat there at Ferney in a pergola in his extensive gardens, wrapped in a fur coat, silk caps, and a woolen hat, writing his witty, cynical attacks on religion and government. One of the characteristics of Voltaire was his love of literary quarrels. He was always picking fights with his friends and allies. Among others, he attacked his optimistic fellow philosopher Gottfried Leibniz's "best of all possible worlds" theory as well as Jean-Jacques Rousseau's ideas about the natural man.

Candide was published anonymously in 1759 in five different countries—not that its anonymity did it any good. Any European intellectual would have suspected it to be the work of Voltaire. Of course there was a scandal; how could there not be? The book exposed the follies of most of the established institutions of the mid-eighteenth century: governments, the Catholic church and its priests, fellow philosophers, politicians, and the army. Voltaire had plenty of material to work with, as much as the era was rife with problems, among them the Seven Years' War. Pitching France and Spain against England, the conflict spread from Europe into North America (where it was known as the French and Indian War).

Naturally *Candide* was banned—which made it all the more popular. But today it is considered one of the major works of French literature and is listed among the one hundred best books of the western canon.

The idea of cultivating a garden in the midst of hard times is not unique to *Candide*. However, Voltaire used the notion as a metaphor or emblem of altruistic reform, one that stood in opposition to the established order of the Church and to blind conservatism. The phrase "cultivate your garden" also appears in letters to his friends. He was not calling for an escape or retreat from the horrors of reality; rather, he was employing the phrase as a philosophical statement of resistance.

That is the point I was trying to make with my garden, although I didn't think of it that way until I got older and the world got worse. The garden was, for me, a literal sanctuary, in the earliest definition of the word. Originally, a *sanctuary* was a section of a Roman Catholic church, one protected from the secular world and its inequities. If you could make it to the sanctuary, you would not be harmed—theoretically. This did not always hold true in real life, as when the toadies of King Henry II killed Thomas à Becket in front of the altar of Canterbury Cathedral.

The very first gardens were wild, often unique spots, similar to the little woodland I had discovered. These sacred groves or hidden vales and hollows were inhabited by gods and demigods; they were watered by murmuring streams and the patter of waterfalls, scented by flowering trees and shrubs and soothed and freshened by zephyrs, the songs of birds, and the hum of bees. Mortals occasionally stumbled into these mystical spots, sometimes never to return. Or if they did, they left with no memory of their time in the place. History is replete with accounts of such sites.

One of the best known is the sacred grove above Lake Nemi in the Alban Hills outside Rome, dedicated to Diana, the Roman goddess of the moon and the wild. Here is where the tree with the golden bough of Roman mythology ostensibly grew. Once cut, according to the myth, the bough will grow again.

According to legend, the grove was underlain by spring wild-flowers and set in a hollow above the lake. I stopped by Nemi once while I was researching Italian gardens, hoping to try to find the actual grove, but it was winter and uncommonly cold, so I gave up. But I could sense the mystery of the site and could understand why it evoked a mythic identity, sitting as it does above Nemi's wild shores.

At some point in recorded time, cultures around the world began to re-create these forested sanctuaries. The earliest of these reshaped environments were actually more like parks or elaborate summer grounds. In Sumer, outside the cities of Uruk and Ur, the nobility reworked the natural landscape into an early version of a zoological garden watered by canals and fountains, planted with fruit-bearing trees and beds of flowering shrubs and herbaceous plants. Semi-tamed local wild animals such as gazelles, wild sheep, and onagers grazed there; nesting birds were encouraged; and in these refreshing environments the urban nobility spent their summers.

These early Mesopotamian gardens were thought to be the origin of later idealized sanctuaries—the Garden of Eden; King Arthur's island sanctuary, Avalon; and, later in history, the Vale of Tempe, in Tivoli, outside Rome. All of these sites were said to be fashioned out of earlier untamed lands. There are references to such places throughout world literature. For example, when Odysseus washed up on the island of King Alcinous he found a garden of tall flowering trees; pear, pomegranate, and apple trees ripe with bright fruit; and sweet figs and olives. Near Emperor Hadrian's garden in Tivoli there is still a sacred vale with falling waters and forested steep hills surmounted by a temple devoted to a prophetess known as the Tiburtine Sybil. Another florid Roman garden is depicted on the walls of a dining hall created by Queen Livia. People honored such hallowed grounds as sanctuaries; they left offerings and even gathered there on certain astronomical dates such as the solstice or during certain configurations of the planets and stars.

With the development of monotheism, the sacred aspect of nature diminished, and people and nature were split apart. In the Judeo-Christian tradition, the male God places man above and apart from the natural world and instructs his people to subdue and multiply and dominate nature. Events in the Garden of Eden sum it up. All is well in the natural world when God creates Adam and puts him into the garden with Eve. But then an evil snake appears and offers the forbidden fruit to Eve, and paradise is lost.

The choice of a snake as messenger was no accident. In earlier pagan chthonic cultures, the snake was sacred and was honored.

This, by the way, did not necessarily mean that the pagan religions were any better at preserving nature than the monotheistic faiths have been. Save for the early hunter-gatherer animistic cultures, pagans killed animals with abandon (sometimes en masse) for no reason other than to keep their gods happy.

But all that is past, for better or worse, and here I am in a garden, an irreligious groundskeeper, planting new beds of roses against a questionable future, laying out beds of annual flowers, and siting in the rose-scented morning sun thinking about nothing or planning new garden beds so that I can grow more plants and thereby, I hope, offer more food and shelter to my fellow travelers, the plants and the animals.

But, as the Buddha came to realize while meditating under a bo tree in the forest, all things are ephemeral.

CHAPTER FOUR

THE PRINCESS OF MONACO ROSE

There was a massive flood late that spring in Nigeria. A vast area of crucial farmland was covered with water, more than 2,000 people were injured, more than six hundred succumbed, and whole villages were destroyed. The UN humanitarian coordinator for the country blamed climate change for this uncharacteristic flood.

The effects of weather events of this sort are causing economic ruin and threatening famine all across Africa, since much of the continent is largely dependent on agriculture. The same is true in Pakistan and Bangladesh; floods and droughts are followed by famine, and uprooted environmental refugees are forced to flee to nearby countries.

After I read the accounts of the horrible Nigerian flood, I went out and bought another rosebush.

The elder owner of one of the Scratch Flat farms has a large family, and several of his children have farms and garden centers all along the Great Road, where Scratch Flat is located. One has a good collection of well-labeled rosebushes complete with notes on their histories. The nursery is my main source of memorial plants.

That day I happened to find there a Princess of Monaco rose, a variety developed in the 1980s by the Meilland family to honor Princess Grace.

There's nothing unusual in this habit of planting memorial roses; gardeners have been using the flower for ceremonial purposes ever since roses made their appearance in Greek and Roman gardens. By the Middle Ages and the Renaissance, they were a standard element of garden designs.

Renaissance gardeners in Italy based their garden ideas on what they saw in the grounds around classical temples. They had accessible models everywhere in the form of Roman ruins, and their gardens often included Roman statuary, loggias, porticos, colonnades, courtyards, and marble stairs leading to temple replicas.

In time, the ornamental gardens of dynastic families such as the Medici became ever grander, so much so that the whole of Italy was seen as the world's great garden. The famous gardens of Tivoli, Frascati, Villa Lante, and the Medici's Tuscan gardens became the models for French and English gardens of the period. Thomas Evelyn and other English designers brought Italian garden plans to England, and by 1575 Elizabethan gardens included loggias, fountains, terraces, labyrinths, and many roses.

In some of these large gardens, designers would leave sections of what the English called "the wilderness," which they based on the Italian idea of *bosco*—that is, land left wild as a monument to the primal forests uprooted by the construction of towns, cities, and controlled gardens.

The classic and most elaborate of the French gardens is Louis XIV's Versailles, one of the most famous in the world, along with the Alhambra in Granada and the Villa d'Este in Tivoli. But one of the lesser known, and no less influential, is Voltaire's Les Délices. He laid out his grounds close to the chateau, planting flowers and herbs as well as apple, peach, and pear trees. He prepared beds of asparagus and artichokes and planted oaks and lindens, elms and maples, as well as cultivated fields of grain and grapevines. And here, in 1759, on a green-hedged sheltered bench in an isolated corner of the garden, he composed his controversial novel *Candide*.

The larger, more formal European contemporary gardens were statements of power and wealth, but there were smaller estate gardens throughout Europe that were used as sanctuaries from the chaos of the urban streets and the *sturm und drang* of daily life.

The wealthier landholders of Virginia had more ambitious ideas. They began modeling their plantations on the designs of famous English garden architects such as Capability Brown, who went so far as to remake the entire contours of land to fit his garden plans. Some of Brown's ideas were incorporated into the garden around Thomas Jefferson's neoclassical house Monticello, just outside of Charlottesville. Jefferson laid out the designs for the property in 1770 and began work on the gardens just as the house was nearing completion. The house itself was based on the ideas of the Italian architect Palladio, but the gardens were influenced by Jefferson's extensive readings in architecture and landscape design. What emerged, according to some historians, was a model of Jefferson's own complex mind.

He started with a gloomy shaded graveyard, then planted a two-acre vegetable garden on a terraced slope on the southeast side of the little mountain (the *monticello*), where the plantation is located. Here he experimented with some three hundred varieties of vegetables and flowers. He constructed a pavilion, which he intended to use as a reading room, and a belvedere where visitors could take in the view of the rolling hills below. He established groves of ornamental trees and an orchard and a berry garden. Closer to the main house, he planted more flowerbeds and constructed a temple surrounded by jasmine, honeysuckle, and sweetbriar. He even planned, as in the early gardens of Sumer, a zoological park to be populated by local species of birds and mammals, including an elk—which he designated as the monarch of the woods—and a buffalo.

To put things into context, I should note that this entire estate, with its acres of ornamental gardens, orchards, and experimental vegetable gardens, as well as the fine, well-crafted architectural details of the buildings, was cleared, planted, cultivated, harvested, and built by enslaved Africans, who, back in their homelands, had been no strangers to craftsmanship or cultivation. This seems to be a characteristic of the grand American gardens and farms. An underclass of enslaved Africans and impoverished immigrants from places as far-flung as China and southern Europe effectively built the infrastructure of the United States. Thoreau, as usual, had something to say about this. He wrote that the rail line that ran past Walden was built on the backs of Irishmen. Were he with us now, he might say the same of the Chinese workers who built the transcontinental railroad. The tradition carries on in our time. What would happen to the American agricultural economy without immigrants?

In any case, thus established, and with the help of enslaved Africans and immigrants, Americans began a long history of garden design, mostly based on late eighteenth- and early nineteenth-century European models. Then, in 1839, Andrew Jackson Downing, a man of humble origins from a small Hudson River town, strode onto the stage and became the most influential house and garden designer of the antebellum decades.

Downing dismissed what he called the ancient style, by which he meant the formal, highly structured Italianate designs, in favor of a naturalistic garden style that attempted to imitate wild nature. For him, this was made easy inasmuch as his wealthy patrons had estates with scenic views of the Hudson River. His design ideas were also helped along by the groundbreaking local renderings created by the landscape painters of the Hudson River

School, such as Thomas Cole and Frederick Church, both of whom had houses on the riverbank.

Downing's gardens included preserved sections of woodlands, and he created paths and vistas to nearby scenic areas and designed houses in the Victorian and Italianate styles that were framed by native trees such as willow and oak and interspersed with sections of fruit trees and fruit-bearing shrubs.

One of his house plans included what he termed a cottage, designed in the rural Gothic style, set within verdant lawns and borders of lilacs, azaleas, and boxwood, mixed with firs, larches, oaks, and arborvitae. The grounds beyond had walkways through green trees and parterres, all of this laid out in a picturesque manner along the front and sides of the house, which was set back from a country road and had a long carriageway leading up to the entrance. The house itself had a high peaked gable center fronted by wide airy porches and two wings, left and right.

In the library of the caretaker's house for what is now the Polly Hill Arboretum, I first came across books by Downing. For me, part of their attraction was his rich nineteenth-century writing style. Nothing was spare; everything was beset with lush verdant greenery, scented bocages, florid parterres, scenic vistas of fields and sheep meadows rolling smoothly down to the sparkling waters of the River Hudson.

But I also was attracted to his house styles: cottages in the "pointed Tudor style," an "ornamental farmhouse," a "cottage villa in the bracketed model," and, one of my favorites, "a cottage for a country clergyman." These so-called cottages, by the way, were not two-room peasant dwellings. They were four- or five- bedroom piles, with verandas, large dining rooms, parlors, libraries, and music rooms.

All of this implied an Old World quietude, a place set apart from the madding crowd, and even though I was only about twenty-two years old, the Downing models haunted me.

When I first began laying out my own gardens, I was living more or less off the grid on the Scratch Flat property in a ten-by-fifteen-foot Thoreauvian cabin without electricity or running water. So when I decided to build a real house, I chose to base the design on one of Andrew Jackson Downing's rural Gothic cottages. (Ironically, my other hero, Thoreau, complained somewhere in his journals about Downing, his contemporary, because he designed "earth-colored" houses.)

While I was living in the cabin, I planted a small garden near a hickory grove, just east of that structure. Then, after I built the main house based on Downing's plans, I started laying out garden plots according to Downing's theories, which included winding pathways, groves of native trees, and flowerbeds and shrub borders based on his concept of a garden as "nature perfected." Eventually, influenced by the Renaissance gardens I had visited in Europe (I was researching a book on Italian gardens at the time), I ended up with a garden modeled on earlier, more formal layouts.

Shortly after the house was built and the rooms and allées of the garden were in place, my oldest brother, James, an artist with a sharp sense of the absurd, decided that the property needed a name, given its links to old-fashioned estates and villas. He said I should call it the Vicarage Garden because he thought the house looked like a place where a retired old vicar might live. (In fact, I had included some design elements from Downing's "cottage for a country clergyman" when I'd sketched out my plan.) I liked the idea but thought it might be more appropriate to call the grounds "Sunny Bank Gardens," which was the name of the former farm that was in operation on this land until the 1950s.

By any name, however, the garden was a decidedly amateur affair. I had no professional assistance. Still, over the course of several growing seasons of haphazard planting experiments and changing layouts, I finally came up with a design based on what is called the *patte d'oie*, or goosefoot, a seventeenth-century French and English garden plan.

Basically, I planted a circular lawn—the footpad of the goose—just off the back porch. Then, spreading out from this footpad, I laid out five paths leading to a series of garden rooms along the southwestern wall, with a teahouse as the focal point, an important element in the designs of Capability Brown.

In the beginning, the garden was a chaos of struggling plants, weeds, invasive vines and shrubs, messy lawns, weedy flowerbeds, too much pachysandra and periwinkle and English ivy. But slowly, after about seven years, it began to age well.

The place was made all the more intriguing (at least in my view) by the new structures I was building on the grounds. The teahouse was first, but at one point I salvaged a narrow antique shed and placed it at the end of an allée of small ornamental trees and shrubs that ran down to the northwest wall of the property. Then, tired of the constant, seemingly daily winter work of digging cars out of snow and ice, I built a carriage-house garage in the Gothic style, with ornate leaded glass windows and cornices I had rescued from a nearby demolished house.

I also began digging out small fish and frog ponds around the property. And finally, struck with an increasing desire to see greenery and to smell vegetation in the cruel midwinters of the region, I built a small greenhouse and later a larger hoop house.

None of these, neither the house nor the garden nor the outbuilding designs, were separated from nature. In fact, they were designed to let nature in. My influences for this layout were ultimately based on the designs of both Downing's house and garden plans and Renaissance gardens. The villas of Italy, and later those of France and England, had wide terraces that gave onto the formal gardens via welcoming steps down to paths and allées and parterres. Downing's designs were also tied to nature, with tall windows and verandahs opening into the landscapes on all sides of the houses and, in the riverbank estates, favoring views to the Hudson.

During his era, American householders were beginning to bring potted plants indoors, quite literally inviting nature in. They also began to bring in an abstracted version of nature in the form of Oriental rugs, which were themselves based on nature, with floral designs of leaves and rosettes, vines and stylized animals.

The careful work of laying out a house and a garden on a tract of empty land is not just an idle folly. There is a strong environmental element to such a design, and there always has been.

Traditional house plans of the past were built in relation to local climate conditions, including seasonal wind direction and the access to sunlight or, in contrast, protection from the hot sun. For example, traditional Mediterranean cultures, including the Greeks and the Romans, arranged their dwellings according to the angle of the sun so that in the morning the living quarters would catch the warming rays but later in the day protect the residents from the blast of afternoon heat. Designs of this sort occur in most regions, especially in deserts, where the nights are cold and the days are very hot.

One of the critical environmental influences on Chinese architecture was (and still is in some projects) the concept of *feng shui*, usually translated as "wind-water" but more generally considered a harmonic alignment with the local environment. The tradition includes larger aspects of the local geography and topography but also astronomical consideration and the flow of "cosmic energy," or *Qi*, a life force that moves through buildings and even the human body. Feng shui was also considered while laying out the designs of traditional Chinese gardens.

When siting a house and garden in Japan, designers emphasize what is known as the borrowed view—that is, placement of the house and garden in such a way to offer pleasing views of the land or waters beyond the property lines.

I laid out my own house with an eye to the position of the sun and the local weather. The Vicarage (so-called) has a lot of windows, but very few face the northwest, which is where the cold winter winds blast in. The many-windowed southeast side gets the most sun, though this doesn't always work in fickle New England, where, as the old Yankees used to say, if you don't like the weather, just wait three days.

Northeasters bring the worst weather. And although the southeast sections get the most sun, the warm rains of summer often come in from that quarter. But on some days everything is topsy-turvy. Years ago a freakish, almost hurricane-force gale swept in from the usually benign southwest and took out a number of pine trees on the northwestern side of the farmland beyond my back wall.

The architects of the big river houses on the eastern shore of Chesapeake Bay designed their structures with two large double-hung doors, one facing the rivers and the other facing inland, so the cooling breezes off the river could sweep through the interior. This was common in the designs of southern architects before the advent of air conditioning. My family's old townhouse on the eastern shore of Maryland had wide porches set with rocking chairs and upstairs sleeping porches that were refreshed by the cool night air. Family and friends would gather every evening on the porch to gossip—mostly about horses, dogs, and eccentric neighbors, in my recollection.

By contrast, the early saltbox designs of country houses in New England face their low sloping roofs northwest, toward the winter winds, and the flat high side of the house southeast, toward the sun. Unfortunately, contemporary architects seem oblivious to environment. One wonders what manner of architect ever thought to design the flat-roofed structures of desert people in northern regions where heavy snows can pile up.

(I shouldn't say this in writing for fear of offending my architect friends, and I am perhaps out of my field, but the designs of Frank Lloyd Wright and Walter Gropius and the Bauhaus group, which famously promoted union with the natural world, may have a lot of glass-plate windows, but you can't open them to let in the fresh air, and some are set so high you can't see outside anyway. Furthermore, the architects seem to have hidden away the front doors and purposely obliterated the welcoming entrance steps so common in the old Italianate villas and in Downing's country houses and, for that matter, in houses the world 'round.)

House design and location have a lot to do with climate change. Traditional architecture took advantage of the local weather conditions, but things shifted after the invention of air conditioning. Even though most contemporary houses still have windows, the owners rarely seem to open them, not even on fresh summer mornings.

There is a large suburban development just beyond Scratch Flat that I often walk through to get to the eastern bank of Beaver Brook. There are about fifty houses in the development, and in every season, including the moderate weather conditions of spring and fall, the houses are locked up tight. Rarely do I see an open window. Not only that, I don't see many people in their well-tended yards save for the workers of landscape companies (and they are equipped with hideous loud mowers, leaf blowers, and pesticide applicators).

The average U.S. household produces 7.5 tons of carbon dioxide a year, which, because of the designs of contemporary air-conditioned houses, collectively releases a significant amount of greenhouse gases each year. I don't know whether any engineering studies have determined how much fossil fuel an earlier house without air conditioning required, but, at least in the north, the only season where the owners would have closed themselves in would have been winter. The windows were open in every other season.

The point is, as far as housing is concerned, we have created an energy-consuming monster, and now we are paying for it.

Over the years, the Vicarage garden began to attract some notice. After all, very few garden designers in our time use seventeenth-century landscaping ideas, and this was, to say the least, an eccentric experiment, coupled with an equally offbeat house and outbuildings that look as if they were built in the 1850s.

(It could have been even more exotic if I had followed the advice of my brother, who was always coming up with new ideas for what I should do to enhance the property. He designed, for example, a pony-cart trail that would run along the outer walls as well as a temple to Diana, one of the standards of eighteenth-century English estate gardens. Then he wanted me to build a long reflective pool in front of the teahouse. He drew plans for an Italianate *collombella*, or pigeonery, where I would raise squabs. And finally, since there were four children living on the properties, he wanted me to build a *garçonnière*, mimicking a nineteenth-century New Orleans tradition of constructing a separate house where the children would live.)

After the house and garden had time to age, word got around within the community, and a few stories appeared in local papers and garden journals, then articles in the *Boston Globe*, the *New York Times*, and *Orion* magazine. The garden was featured in the July–August 2014 issue of *Horticulture* magazine, and students in a photography course at Harvard University, led by the landscape photographer Frank Gohlke, descended on the garden one summer afternoon and scoured the property for subject matter. Another photographer friend even produced a book of photographs of the garden.

I began to get requests for garden tours, and, in time, we began having annual garden parties, sometimes to celebrate the publication of a new book, sometimes to celebrate peonies. (The garden parties were always held when the peonies were in full flower.)

Among the various friends and allies who would come to the house was a scholarly older man named Robert Rosen who held advanced degrees in two different fields and had ended up working in a third. Robert began reading when he was four and didn't quit. He was still doing research in new fields of study at the age of seventy-nine. He was no polymath; he knew little of science or math or engineering, but his expertise in other subjects ranged through the entire body of eastern and western thought. He could read ancient Greek, Hebrew, and Latin and had a familiarity with Sanskrit, and he was interested in the lost languages of Native Americans and their predecessors, the native Siberians. He was also well read in theology, ontology, and eschatology, and, to a degree, ecology (from the ancient Greek, as he pointed out, meaning "the study of the house").

He was an energetic, enthusiastic old teacher, according to one of his students. He was full of opinions, some of them controversial in our time—a supporter of old-fashioned western cultures, a reader of obscure histories and antique literatures in forgotten languages. He could hold forth tirelessly (some thought tediously) on almost any subject, even those he didn't know much about.

Robert had so many theories about such a wide variety of subjects that I'm not sure he could have been an authority on any of them. But I once asked him about the origin of the Amazons (a subject I was interested in because of my own interest in the domestication of horses, not exactly a field of study for most people), whereupon he delivered an informed lecture on the wild tarpan horse and the Botai cultures of the steppe tribes of Scythia. He told me that these tribes used to hunt horses, then began herding them, and then, in time, learned to ride on their backs. He claimed that this shift allowed the formerly isolated tribal groups to spread their language throughout the steppes and even further afield. He claimed that their language is the root of all Indo-European languages. He went on to explain that their cultures were traditionally led by violent horse queens, warrior women who became known as the Amazons. Even the Greeks feared them.

(And I thought I could stump him with my Amazon question!)

To say Dr. Rosen played Dr. Pangloss to my Candide would not be entirely accurate. He was far too cynical, and in fact he was a little more like Voltaire than like his character: he was always writing ripostes to articles in the *New York Review of Books*, for example, and getting into literary arguments with fellow academics (as did Voltaire). But I did like checking my sources with him, and he did like holding forth.

Still, at some point, having discussed Voltaire with him, I jokingly began to refer to him as Dr. Pangloss. Scholarly though he was, Robert was not lacking in humor, even when he was the subject of derision, and he came to enjoy the absurdity.

It is said that a gardener's work is never done. For my part, I carried on over the years, season by season, year by year, designing new beds, planting new trees, digging and shearing and pruning, planting and weeding, and, in winter, getting exercise by shoveling snow to clear the serpentine paths that wind through the garden. And all the while, beyond the garden walls, the world was getting worse.

The 1,200-year drought in the American Midwest showed no signs of ending. A huge portion of the Ross ice shelf, the size of Connecticut, broke off. Worldwide, glaciers continued to melt. Wildfire seasons in the West became a constant and grew larger and larger and ever more destructive because of the drought. Islands from the Chesapeake Bay to the South Pacific began to sink beneath the waters. Floods overwhelmed vast sections of low-lying nations and ever so slowly the levels of carbon dioxide continued to rise.

Sic transit gloria mundi, as Dr. Pangloss might say.

CHAPTER FIVE

THE WHITE ROSE

Years ago, at an award ceremony, I met a tall gentlemanly figure who was the keynote speaker for the event. When I arrived at the ceremony, the tall man was in conversation with a woman about his most recent book. I joined the group and asked him what his book was about. He looked down at me as if in surprise.

"Why, ants, of course," he said.

That was my first meeting with the biologist and author E. O. Wilson.

Over the years I found myself often doing research at Harvard's Museum of Comparative Zoology where E. O. (as he was known to the cognoscenti) had his office, and I would run into him from time to time and chat. Later, I got to know him socially and was invited to his eightieth birthday and other events. But the longest talk I had with him took place when I happened to meet him while waiting at Boston's Logan Airport for a flight that had been delayed.

I was familiar with his books, especially his ambitious 2017 work *Half-Earth*, which had just come out. His argument in that book was that, rather than save bits and pieces of critical habitat, we should work to save the few vast ecosystems that are still intact to preserve the Earth's biodiversity. Large sections of the Amazon and Congo rainforests are still unspoiled, as are sections of southern Chile and mountainous sections of New Guinea. From these plots of wild land and natural systems, we would be able to at least hold at bay the current decline. Wilson was in his early eighties at the time, and it was refreshing to hear such a positive solution from a man who had spent most of his life fighting an apparently losing battle to conserve wild nature.

Major extinctions of established species are nothing rare; five different extinctions have taken place over the Earth's 4.5 billion-year history. This current Anthropocene era is the latest. The first, the Ordovician extinction, occurred about 440 million years ago and wiped out nearly 85 percent of the Earth's newly evolved species. Then, about 360 million years ago, there was another die-off, the Devonian extinction, which was spread over about 10 million years and caused the extinction of nearly 80 percent of species. Following this there was a period that marked the successful spread of plant life on dry land, an event that, conveniently for the aborning animal life, consumed vast amounts of carbon dioxide.

But of all these disastrous epochs and eras, the one we should care about most is in the Permian-Triassic extinction, which began about 252 million years ago, an era known to geologists and paleontologists as the Great Dying because so much of the struggling and seemingly tentative attempts of life on the planet were killed. This massive death created what is known as the Carboniferous Period, which is the source of our current problems with global warming.

The land plants of this period were nonflowering species: huge ferns the size of trees, immense club mosses, and hundred-foot-tall horsetails. Flowering plants, trees, shrubs, grasses, and forbs would not appear for more than 100 million years.

But like all green plants, the nonflowering plants began absorbing carbon dioxide and giving off oxygen through the process of photosynthesis. Levels of carbon dioxide in those alien times stood at about 4,000 parts per million. But slowly, over millennia, with the evolution of flowering plants, green plants absorbed most of the carbon dioxide, which in turn allowed primordial animals to evolve, including, about 250 million years ago, the dinosaurs.

As it turned out, however, there would be consequences. (There are always consequences, as Dr. Pangloss points out.)

During the Great Dying, all of the stored carbon in the dead plants was absorbed into the dank earth, and the dank earth evolved into peat; and over the eons, the peat dried out and was compressed and became a rock-like substance known as coal.

Oil and gas were formed from the aquatic plants and marine organisms found in the great oceans that covered most of the planet in those distant eras. After these organisms died, they were buried on the ocean floors by sediments and over millions of years were compressed deeper and deeper into the earth and transformed by the Earth's interior heat and intense pressure into petroleum products. At some distant prehistorical point in time, a newly evolved Cro-Magnon individual learned that the black stones that showed up on the surface of the ground in some places would actually burn when exposed to fire, and burn more slowly and with a steadier flame than wood, the usual source of fire. There is archeological evidence that the Chinese were burning coal about 4,000 years ago. Then the Romans discovered the fuel and used it for heating. And then, in the mid-eighteenth century, in what turned out to be a critical event as far as the Earth's atmosphere is concerned, the English began mining coal, which they used to power the dynamos of the Industrial Revolution, which in turn created the foggy, air-polluted cities of the mid-1800s.

In certain regions on Earth, the oil created by dead marine species would bubble to the surface and, as it seeped, would mutate into tar pits and bitumen, or asphalt. Tar was originally used as a sealer in ship building; it was not until the 1850s that this crude oil was refined and used as fuel.

In 1824, Joseph Fourier discovered the presence of a glasslike barrier around the Earth created by carbon dioxide, which can't escape the Earth's atmosphere because it's too heavy. Gravity keeps it from drifting above the lighter atmospheric components, and the heavier carbon molecules don't move as fast, so they can't get out. This means that the smoke released everywhere on the planet, which includes, among other gases, a high amount of carbon dioxide, will dissipate but cannot escape into outer space.

One could argue that our current period of climate change began in China and Sumer, with the development of agriculture. Or we could place it even earlier, since it was at some point in the Neolithic Era that people began using fires for purposes other than cooking.

Fire may also be the key to one of the great mysteries of biological history. What happened to the mega-fauna species of North America during what is known as the Pleistocene extinction?

Just a few thousand years ago, with the retreat of the last glacier, the American continent was populated with giant beavers, immense birds, lemurs, huge maned lions, crocodiles that weighed a ton or more, and the better-known woolly mammoths, mastodons, and saber-toothed tigers. Although paleontologists have not entirely determined what happened to these prehistoric giants, climate change was long a primary theory. But in the 1950s, a paleoecologist at the University of Arizona named Paul Martin pointed out that the extinction of the mega-fauna coincided with the arrival of Siberian hunters who crossed the Bering Straits land bridge about 15,000 years ago.

The American fauna had never before encountered a small upright mammal that walked on two legs, so the animals had no fear. Using efficient fluted spear points, these Paleolithic Era hunters could surround a mammoth and dispatch it. But they may also have used another hunting technique that involved a system of controlled burning in which grasslands were ignited upwind of a herd of mastodons and used to drive the animals over a bluff. This would account for the assemblage of mammoth and mastodon bones archeologists have discovered at the base of cliffs, along with Paleolithic butchering tools.

There is also evidence that the eastern woodland cultures that were active during the past 4,000 years practiced a form of land management that predated the development of agriculture. People would burn off sections of forest to encourage the growth of blueberries, which would in turn attract bears and white-tailed deer. They would then hunt the animals, then mix dried strips of venison with bear fat and blueberries to create pemmican, a staple food used for winter survival.

Human populations were still small in the Neolithic Era, totaling only about 10 million people. (There are more than 7 billion now.) Levels of carbon dioxide in this preagricultural period were about two to three hundred parts per million. But things were about to change.

Roughly 10,000 years ago, in the Mesopotamian river deltas, early cultures began encouraging the annual growth of native wild grains, leading eventually to more and more active control of the annual floods through the creation of a system of dams and canals and the development of storage systems for the collected grains. The practice also led to permanent settlements, which eventually grew into urban centers, the first and greatest of which was Sumer.

Ancient Sumerian tablets, discovered and translated in the 1940s, provide details of agricultural practices for the production of grain, the cities' staple food supply. But the tablets also include details about planting shade trees and growing vegetables, essentially chronicling the first gardens. During this period, the practice of harvesting trees for construction and firewood also began.

One example is documented in the Sumerian *Epic of Gilgamesh*, which was set down around 1200 BCE. In the beginning of the story, Gilgamesh and his beloved friend Enkidu battle a monster named Humbaba, who guards a forest of cedars maintained by the gods. The friends defeat Humbaba and then celebrate their victory by cutting down the sacred forest. And so begins the path toward our current climatic situation.

A single square mile of forest can contain as much as 30,000 metric tons of carbon. Still more is stored in the plants of the understory. When the trees are cut down, much of the carbon is released back into the atmosphere. For example, over a twenty-year period, the development of new agricultural lands can raise the carbon level by about 10 percent. If you consider that statistic alone, you get an idea of how clearing land for the development of agriculture over 6,000 years would naturally increase the amount of carbon stored in the atmosphere.

It used to be said that, before the early agriculturalists of Britain began chopping down the woodlands, a squirrel could travel across the tops of trees from John O'Groat's in the north of England to Land's End in the south. And before the English arrived in North America in the mid-seventeenth century, a squirrel could supposedly make it from the East Coast to the Mississippi River without ever touching ground.

Studies in the 1930s, carried out by researchers at Harvard University's forest, determined that this claim is not accurate, at least not in North American, where there were large unforested sections caused by fires, hurricanes, and beaver meadows. But the fact remains that by 1850 only isolated tracts of woodlots remained in the East, most of the land having been cleared for agriculture, building materials, and heat. New England alone was more than 80 percent cleared. Today, however, the statistics have switched, and the Northeast is currently the second-most-forested section of North America. The first (and oldest) is the great forestland of the Pacific Northwest; most of the forest in the Northeast is second growth.

With such a vast amount of forested land, what would a small patch of forest such as the unspoiled woods in my demesne matter?

But as E. O. Wilson and others have pointed out, what we save now is all we'll ever save. And as it turns out, not all current agricultural systems have to be destructive.

CHAPTER SIX

THE DAMASK ROSE

G iven the fact that human beings have now affected all of the ecosystems on Earth, it seems ironic to say that, done properly, with an eye to local ecology and organic practices, a garden or a farm can be restorative. But as it turns out, the evidence is there. The entomologist Douglass Tallamy has written a series of books that offer information on how to manage a property for the benefit of local species of plants and animals. Essentially, he sees the typical suburban home garden as the last best hope for preserving nature.

His first and most import advice to any householder who is interested in attracting wildlife to a property is to shrink the lawn or get rid of it altogether. In our time, lawns cover most of the managed lands in the United States, more even than agriculture, which gives you a sense of how much habitat would be created by lawn removal alone. But Tallamy goes on to document specific flowering herbaceous plants—trees, shrubs, and native species—to encourage insect populations that are, along with plants, the base of the food chain. One of the theories on the decline of bird species links to the fact that insect populations are dropping. According to studies carried out in Germany, insects have declined by as much as 50 percent, and there is new research indicating that urban light pollution is affecting night-flying insect species and migratory birds.

Years ago, the English entomologist Jennifer Owens, an authority on ichneumon wasps, made the same point; she and Tallamy, used their own property as a living example of what could be done. Owens claimed, on good authority, that all of the habitats in England are threatened, save for one—the home garden. She insisted that a typical household garden, planted organically and ideally using native species, can stabilize and even increase plant and animal diversity. Tallamy also created a living example of a backyard sanctuary, what he came to call a homegrown national park. His house was built on a species-poor hayfield in a neighborhood of typical suburban houses with requisite front lawns. But he and his wife redesigned the grounds and created a landscape with native plants that would attract specific kinds of bees, butterflies, and moths.

My plot, Sunny Bank Garden, offers another example—although when I first laid out the garden I had no idea that I was in any way restoring land. In fact, I had started the project by destroying a natural habitat—although the white-pine forest itself was an artificial monoculture habitat that had grown up after the colonists cut down the original 7,000-year-old hardwood forest of oak, maple, and hickory to make way for farmland.

In the early to mid-twentieth century, after many of the farms in this region were abandoned, the dominant trees that grew back were the sun-loving white pines. But as my before-and-after ecological survey made clear, Sunny Bank Garden increased the biodiversity of this one-and-a-half acre plot from five or six higher species of plants and animals to 2,000—and that is a conservative estimate, according to the Harvard ecologist Richard Forman.

Transforming the traditional American yard into a mini wildlife sanctuary would restore species that favor an edge ecotone—that is, a habitat that favors small mammals, white-tailed deer, foxes, and coyotes as well as many species of birds and a large community of insects and other invertebrates. But that does nothing for many species of deep-forest animals such as bears, bobcats, porcupines, pine martens, moose, and mountain lions. Land trusts and other forest preservation groups need all the support they can get to maintain large sections of local forested lands of this sort.

In the meantime, we still have to eat. Most commercial farmers are not in the business of preserving wildlife. In fact, their profits are often threatened by wildlife. Raccoons and deer (or both, working together) can destroy a sizable portion of an annual harvest of corn, tomatoes, and fruit in a single night. So can woodchucks and squirrels, though the apparent return of coyotes and bobcats to New England helps. (I watched, possibly too close at hand, late one spring as a pack of determined coyotes chased down a fawn in the wooded edge of the farm just over the hill from me.)

The white-tailed deer populations in the East have expanded so much that they have become a pest species. More mountain lions would help—but there are not enough of them currently, and I don't think the residents of the expanding suburbs would appreciate a top-level predator sauntering through their backyards, snatching up a beloved dog or cat as a snack en route to a deer hunt. Coyotes are bad enough: one carried off a friend's Jack Russell terrier, and years ago I saw a huge wolf-sized coyote eyeing my own Jack Russell as a potential dinner. These terriers are egotistical little dogs; most believe they are huge pit bulls and fear nothing. Mine was ready to attack if I had not restrained him. A year later he was badly mauled in a fight with a coyote.

But farming techniques are changing. Organic farms are expanding outside urban areas, and some farmers beyond the suburbs are working with systems to establish a compromise with local pests by planting buffer zones of crops or following the old farmer's proverb for corn planting: "One for the blackbird, one for the crow, one for the woodchuck, and one to grow."

Nevertheless, the carbon sequestration of a typical farm does not equal one hundred acres of forest. Furthermore, simply replanting trees in commercial forestry operations is not enough to counter the rising tides of climate change. What is needed is the preservation of the traditional wildwood.

E. O. Wilson pointed out that restoration of our ecological systems are the natural means to bring the Earth and the air back into balance. There are aspects to ecology that may not be fully understood, such as a recently discovered interrelationship in which fungi make soil nutrients available to various species of trees. The large ecological complexities of a healthy forest may be more effective in controlling carbon dioxide than are plantations of a single species of trees, even if the tree variety is genetically engineered to absorb more carbon dioxide than other species of trees can.

In addition to organic practices, some farmers are now redefining the size of their farms and turning portions of their land over to wild nature. Following a new development called regenerative agriculture, farmers and landowners intentionally allow their fields to grow back into the preagricultural habitats that originally characterized the region.

One successful example is the Knepp Estate, a large farm of some 3,000 acres in West Sussex, England, that has been in the same family for generations and is currently owned by Charlie Burrell and his wife, Isabelle Tree. Burrell inherited the farm from his grandparents; but when he took over, it was already losing money. The soil was heavy with clay and hard to cultivate, and although the family tried to resuscitate it, they went through years of losses. Financially stressed and facing competition from larger commercial agricultural systems, the owners decided to change course. They sold all of their farm equipment to their competitors and then, having eliminated their debts, they allowed most of the farm to return to its natural state.

Suffice to say that the experiment was a great success and the means of achieving that success was complicated and imaginative. After all, what does the catch-all phrase "return to nature" actually mean? What, in fact, *was* the original habitat of northern Europe and Britain?

After its emergence from the ice prison of the late great glacier that covered the northern hemisphere until about 15,000 years ago, Britain had no forests. The lands at that time would have been a little like the contemporary tundra, covered by heath-like plants and barrens of trees. These conditions may have endured for 1,000 years or more, but slowly, as the climate warmed, higher plants began to move in. As the waters of the melting glacier were released, large areas became covered with shallow pools, ponds, and freshwater wetlands populated with the kinds of wetland vegetation that we are familiar with today, such as cotton grass and willow. Eventually, as the lands dried out, trees began to appear. The forest by this time would have been essentially taiga: the vast tracts of spruce and black fir that still cover much of the northern hemisphere, from Scandinavia, Russia, and Siberia all the way across North America. But in the region around Yorkshire and Nottinghamshire, these northern boreal forests slowly gave way to the various species of deciduous trees that make up what's left of the wooded lands that cover Britain in our time: English oaks, beech, birch, and holly, plus patches of giant bracken ferns.

About 12,000 years ago, Cro-Magnon people began to move into the British Isles, which were, at the time, connected to Europe. The tribes were originally nomadic hunters, following the herds of woolly mammoths, reindeer, and giant oxen as well as brown bears, woolly rhinoceri, and other game animals that moved in as soon as the glacier retreated. During the late Stone Age, the hunters began gathering plants, and eventually they settled in small temporary communities. Then, about 6,000 years ago, they began to practice an early version of agriculture. And with this monumental change, from hunting to agricultural use, the ecology of the island began to change again.

Some ecologists and forest historians believe that the introduction of small plots of farmland in the late Neolithic Era may have actually improved the biodiversity of the original deep forests of Europe and the British Isles. The vast dark continent of tree cover was not particularly diverse as far as wildlife was concerned. The aurochs, wild tarpan horses, and other grazing species were not common in this ecosystem. But as the early agriculturists began opening up meadows and fields, their numbers increased (albeit only to be killed off later as human agricultural populations spread over Britain and the Continent).

The rewilding of Knepp Estate was modeled on the environment that existed during this particular period of change: from deep forest to small plots of open land. But given all of the changes that had taken place more recently—most dating back to the Victorian Era after the destruction wrought by the Clearances—returning the land quickly to this state would have required a lot of hard human work and machinery. Instead, the family decided to let animals do the job. They introduced Exmoor ponies, one of Europe's oldest breeds, closely resembling the tarpan horses that roamed the steppes of Eurasia and Britain during the Stone Age. They also brought in Old English longhorn cattle, an ancient breed similar to the extinct aurochocs that, until they were killed off in the seventeenth century, were common in the British Isles. The Tamworth pig replicated the wild boars, which do still exist in England but have no predators and can wreak havoc in forest habitats if their populations expand. The family also introduced roe deer and fallow deer, both native species. In the years following, other original species of birds, mammals, and insects began to move in, including some that are rare in the British Isles, such as the Behstein's bat and the dung beetle. In the now favorable habitat, all of these species began to increase.

The Knepp Estate was not Europe's first successful rewilding project; there was an earlier and much larger restoration in Holland. Known as the Oostvaardersplassen, the twenty-three-square-mile tract is part of a polder, a section of lowland reclaimed from the sea (as is much of the Netherlands). The Dutch ecologist Frans Vera came up with the idea of using grazing animals such as Konik ponies, Heck cattle, greylag geese, and red deer to re-create the late Stone Age environment of Europe. Created in one of the world's most densely populated nations, the project became the most successful rewilding experiment on Earth. Once established, the restoration attracted not only birds that were already living in the region but also some birds and mammals that were not known to still exist in Holland.

The environment created by these restorations is a landscape of trees scattered in open grazing lands, with ponds or lakes spread throughout. Such postmodern ecological experimentation may have antecedents, according to my friend Dr. Pangloss Rosen. He claims that this landscape evokes atavistic memories of the habitat of the earliest hominids: the veldt of southern Africa, a landscape of grasslands offset by clumps of trees. E. O Wilson promulgated this idea in his book *Biophilia*. He pointed out that, even now, when high-powered computer corporations develop an "office park" (a term that does not in any way match my understanding of the word *park*), they surround these corporate headquarters with open grassland, sparse plantings of trees, and buildings that overlook water.

Deep dark forests seem to have the opposite effect, at least in the mind of earlier agricultural western Europeans, who have traditionally seen them as dangerous places populated by demons, witches, and hideous monsters. Even the highly structured Italian gardens of the Renaissance left a little uncultivated land in the form of the bosco, which could be read as a symbol of victory over the dark godless wilderness populated by wolves and leopards, that Dante described at the entrance to Hell.

✿

Just southeast of Scratch Flat, there was a potential rewilding opportunity. Adjacent to a large tract of forested land that was saved from development some twenty-five years ago there was a thriving, privately held apple orchard. In his will, the old farmer wisely decreed that the property could be sold only for agriculture

The orchard was located in a uniquely picturesque and historic landscape, set on a wooded hill that rolls down to a large pond, which provides the nearby town of Concord with water. It is believed to be the site of the Nashobah Plantation, one of sixteen villages of Christianized Indians that the proselytizer John Eliot founded in the 1640s.

After the farmer died, the town took control and officials leased the land to an orchardist; but when he retired, they were unable to find a replacement. So after several years, they decided to privatize the land and sell it to a farmer without restriction.

Not a good move.

Legally the land could only be used for agriculture. But if the farmer who purchased it were to sell, who's to say the new owner would be a farmer? This was prime property, located near a highway, and local land prices were soaring. A wealthy developer could go through the expensive process of revoking the agricultural restriction and create yet another desert of manor-like five-bedroom houses with pesticide-maintained sterile lawns.

The townspeople rose up. No one wanted to allow the orchard to be chopped down and turned over to who knows what; no one wanted to have their view despoiled by mansions. A committee was formed, meetings were held, the debate began, and alternatives were suggested.

I was there, and I proposed an idea: do nothing. Let the orchard return to nature. In other words, designate the preservation of the orchard as a regenerative agricultural project and allow it to return to nature. Old untended orchards are early seral habitats, similar to abandoned fields but even more diverse, and attractive to a larger number of wild local species. For instance, the highly desirable morel mushroom favors orchards. And wild orchards attract a number of birds species, including the increasingly rare brown thrashers and Baltimore orioles and a wide variety of sparrows, warblers, and flycatchers. They also draw mammals such as deer, three species of mice, and their predators—hawks, owls, coyotes, bobcats, fishers—as well as the usual inhabitants of suburban backyards: raccoons, skunks, and possums.

Dr. Pangloss Rosen and professional ecologists point out that the climate-related benefits of the preservation of whole ecosystems—projects like the Knepp Estate and the Oostvaardersplassen—may not be entirely understood, but they are certainly better than those of commercial farmlands. Twenty-three acres of orchards consume ten to twenty tons of carbon dioxide a year, far better for the climate than huge carbon-producing suburban houses.

But my idea went nowhere.

Ours is a democratic town and generally open-minded, but this was, according to the legal authorities, going too far. The official negative decision was based on the exact meaning of the word *agriculture*, which is defined as "using the land for growing food."

My argument in response cited Thoreau's essay "Wild Apples." I pointed out that the old untended orchards do indeed provide food, not necessarily for people but for many species of local herbivores, including deer and bears.

That argument, too, fell flat.

Toward the middle of April 2022, there came one of those days when the young sun was halfway toward the summer solstice. Tulips and forsythia and magnolias and cherries and crabapples were in bloom; the songs and calls of flickers, wrens, robins, and early warblers were filling the air; and the very earth spoke of life everlasting.

I went out to my newly created morning garden on the southwestern side of the grounds, lifted my face to the sun, closed my eyes, and fell into one of those reveries in which all that was real was the here-and-now: eternal rebirth, life breaking forth from a dead land, forever and ever.

There were no wars, no murderous bands of rebel armies storming the walls of the garden, no floods, or droughts, only the warmth of the sun, the scent of fresh vegetation, and the surrounding peace of a garden, detached from the world.

But one should never read the news.

The next day I learned that a fast-moving firestorm in Greece, caused by drought, had driven local villagers into the sea to escape the fire. Many could not outrun the flames.

Such are the contradictions of our time. Is this, as the old fool Dr. Pangloss repeatedly tells Candide, the best of all possible worlds? Or are these decades, as the ecologists and climatologists and the paleontologists tell us, the worst of all possible worlds?

We may not know the answer until 2050.

CHAPTER SEVEN

THE NEW DAWN ROSE

There were a number of record-breaking climatic disasters that spring and early summer. The Colorado River hydroelectric dams that created Lake Mead and Lake Powell had been slowly drying for years, leaving docks and boat launches far from the shoreline. By 2022, both lakes were half empty. And because of faulty calculations when they were first planned, scientists believe they will never fill again, an error that jeopardizes the water supply for nearly 22 million people in the dams' lower basin.

The region was having its worst drought in 1,200 years. Engineers calculated that if it were to continue, the waters would drop below the power pool that turns the turbines and cut off electricity for nearly 6 million people living downstream.

Quite by accident, at the time, I happened to be rereading Edward Abbey's 1967 wilderness paean *Desert Solitaire*, his account of a year spent working at Arches National Park. In it, he describes a time when he and two companions made one of the last raft trips down the free-running Colorado River through the rapids of Glen Canyon.

I still remember the controversial environmental battle to halt construction of the Glen Canyon Dam and save the haunting, almost otherworldly beauty of the river in this section of its long journey to the Gulf of California in Mexico. In our time, because of climate change and overuse, the river no long manages to reach the gulf. It simply dries up in the desert sands before it gets there.

The dam was already under construction when Abbey ran the river, and his book offers a lament for its loss. It describes the deep, narrow, shadowed canyon, which in places nearly blocks the sky. As Abbey wrote, all along the route, the sad descending call of the canyon wrens joined his lamentation like a fugue.

Soon gone forever, all this.

Or is it?

I imagined the great irony of the fact that, if the waters continue to drop, at some point in the future the canyon will reappear, albeit without a Colorado River.

The current massive drought is causing water levels to drop throughout the region and is also threatening one of its most ecologically significant features: the largest of the western water bodies, Utah's Great Salt Lake.

In the past, under normal conditions, the waters of the lake would shrink by several feet every year in late summer but then would replenish itself each spring as the snowpack in the nearby mountain ranges melted. By 2022, thanks to the prolonged drought and a burgeoning population's increasing demand for water, the lake was at historic lows and still shrinking.

As a result of climate change and higher temperatures, the snowpack is released as vapor rather than water, which normally runs down to the rivers and streams that feed the lake. The overall ecological effects of this are devastating.

The lake is famous for its brine shrimp and its huge population of brine flies. All told, there are more than three hundred resident species of birds that depend on the ecosystems associated with the lake. Furthermore, each year in autumn and spring vast numbers of migratory birds pass through, including eared grebes, white pelicans, phalaropes, and plovers—nearly 10 million birds in all descend on the lake to feed on the shrimp and the flies. Now, with the decrease in brine fly and shrimp populations because of the drying lakebed, the numbers of migratory birds have fallen dramatically.

Meanwhile, there are other consequences of the drought. The dried-out lake bottom contains high levels of arsenic. And as the bottomland dries up, windstorms carry toxic, arsenic-laden dust into the local human communities. Short-term effects include coughing, sore throats, and the like, but smaller particles known as PM 25 can lodge in the lungs and cause more serious respiratory symptoms and even lead to cardiovascular disease.

There is even worse news. A UN report indicates that, if the drought continues, the lake may dry up completely in five years.

The same drought diminished the waters of the Mississippi. In the summer of 2022, barges carrying cargoes intended for downstream ports were stuck upstream because the water wasn't deep enough for them to navigate. The wreck of an ancient ferryboat surfaced and, most interestingly, archeologists discovered on the dried-out river bottom the tooth and jaw of an extinct American lion, a species that died out some 11,000 years ago during the Pleistocene Era.

Mega-droughts of this sort were also happening elsewhere in the world. Reservoirs in Spain were dropping, revealing, among other things, Neolithic stone circles. And high winds in Spain and Portugal, as is often the case during droughts, triggered devastating brush and forest fires.

These global droughts could affect nearly 75 percent of the world population in the next few decades, according to a UN report. This means that close to 3 billion people will face water shortages.

Associated with droughts is another increasing environmental threat—wildfires. New Mexico, along with other southwestern states and California, seems to have developed a fifth season, one featuring massive out-of-control wildfires that ignite in summer and spread over thousands of acres, in some cases burning down entire villages.

Historically, forest fires were started by lightning strikes. Though they would spread, they would eventually burn out and, in time, the forest would grow back. In some areas, fire has actually shaped the ecological makeup of a region, including unique species that have adapted to regular burns. The Kirtland's warbler, for example, is a tiny slate-colored bird with a bright yellow breast that nests in the jack-pine thickets of Michigan's Lower Peninsula. Jack pines depend on fire: the extreme heat bursts the cones of older trees and spreads seeds that develop into new growth. Kirtland's warblers also rely on those seeds. But because of human activity—mainly timber operations, which began in the 1800s, and later fire-prevention operations—the cycle was broken, and the bird was driven almost to extinction. At one point in the 1970s there were fewer than four hundred in existence. Ornithologists alerted conservation agencies, and state and federal regulations managed to halt the decline. The warbler population began to recover, and the bird is no longer on the endangered species list.

In reality, forest fires were not a problem until people began settling in affected areas. That spring, when the tulips and daffodils in my garden were springing up from the warmed, moist earth and the red maples were flaming with their opening flowers, the deadliest fire in New Mexico's history broke out and began to spread across the state. The largest of several separate blazes covered five hundred square miles, and two villages were entirely burned out.

This was just one of eight major fires that burned that summer, destroying more than 1,130 square miles of forest. It was a terrible year, hell fires everywhere, yet climatologists claim, as with so many of these disasters, that this is the new norm.

And while the West burns, I am in the garden, planting new beds of roses, laying out annual flowerbeds, and sitting in the morning sun beside the greenhouse, daydreaming about creating a new garden allée over on the southwestern side of the property, where a large boulder marks the graves of the various cats and dogs that once lived on this land alongside the populations of squirrels, chipmunks, foxes, raccoons, possums, nesting birds, snakes, salamanders, frogs, and uncountable species of insects that shared their space.

To memorialize the New Mexico fire, which was very much weighing on my mind, I went out and bought more roses. In fact, I bought one of the most resilient of varieties, a rosa rugosa, which can stand all manner of abuse and still thrive. But a week or two after I planted it, it started to lose leaves. Even though I watered it religiously, it could not cope with two weeks of hot sun, with no rain in sight. So I transplanted it to an older bed, where the soil was deeper, and it slowly recovered, offering yet another metaphorical statement on climate change. The world has no master gardener to keep things in order.

CHAPTER EIGHT

THE REINE VICTORIA ROSE

B efore the drought set in, the heavy rains that brought on the floods that spring fostered a fertile early summer. Things went well in the garden, but the climate-related disasters seemed to be building toward a record-breaking year on all fronts— historic wildfires, the worst droughts in memory, terrible floods in Bangladesh, a third of Pakistan under water. Furthermore, new studies indicated that the Arctic is warming four times faster than climatologists thought it would, which means that predicted droughts, floods, and rising sea levels will be coming sooner than expected.

In his later years, my version of Voltaire's Dr. Pangloss, the good Dr. Rosen, moved to a retirement home near me. There, he carried on with his research, and I continued to visit him whenever I needed to get perspective on my subject of the day, which at that time was climate change. Weather permitting, we would sit out on a patio on the southeastern side of his building, overlooking the Sudbury River.

I joined him there on a late spring day that year while I was deep into climate change studies, and we started talking about the effect it would have on cultural traditions.

Of course, Dr. Pangloss Rosen knew about Voltaire's garden metaphor (how could he not?) and about my own view of gardens as both ecological and spiritual sanctuaries. But he said Voltaire was dealing with another issue in *Candide*.

"The book was a sendup," he pointed out. "Voltaire was specifically ridiculing the popular philosophy of Gottfried Leibniz, who argued that, since God was perfect, nothing he did on Earth could be imperfect, so anything that happens must therefore be, *en fin*, 'the best of all possible worlds,' as the old fool keeps telling Candide." This was, after all, the Enlightenment, a period of cultural history in Europe when philosophers and writers were questioning the nature of the deity (not to mention reality, the meaning of life, and other light matters).

He also pointed out, citing past and present critics, as he often did in his pedantic manner, that Voltaire's decision to have his characters stay put and cultivate their garden could be read as a statement on the importance of individual responsibility, somewhat similar to the Buddhist statement of right action, or even the existential idea of personal action as a statement of purpose—what you do is what you will the world to do, or something like that. He said the ending stirred up a lot of interpretation in French literary discussion that went on over the years; still goes on, in fact. Among other readers, he said, Gustave Flaubert, who claimed to have read *Candide* twenty times, thought the ending was brilliant. Others disagreed. (Of course—Pangloss said—it's what French critics do.)

He pointed out that Voltaire was himself a gardener, who supported other gardeners and the idea of a garden as opposed to a pragmatic agricultural field or a commercial farm. He told me there was an early interpretation of *Candide* suggesting that Voltaire's garden metaphor meant we should just take advantage of any opportunity to get ahead: "have an eye to the main chance," as it was phrased, which is not illogical, given Voltaire's business acumen.

I told my Dr. Pangloss about my own interpretation as it relates to climate change—as a symbolic statement of what the world should do in the face of things. Just make a statement by creating a tiny part of the planet into a biological model that not only absorbs carbon dioxide but also encourages biodiversity and offers solace, even if that statement is only a geranium on a windowsill.

He said Voltaire probably wouldn't agree with that, though it does in some ways reflect the innocence of the young hero, Candide. "All those hideous things that happened to Candide were based on real events of his time—60,000 people dead in the Lisbon earthquake, the abject cruelty of slavery, the autos da fé, the absurdities of war. All those things are still relevant, still happening."

I agreed, but I said now we're in a situation Voltaire could never have imagined—a threat to Earth itself.

"I think Voltaire could have imagined it," he said. "There were religious fanatics marching through the streets, preaching the end of the world in his time as well as ours."

"Yes, but this time it's real," I said

Researching the background of *Candide*, I had read somewhere that Voltaire, for all of his liberal rants, was an antisemite, an odd stance, given his rhetoric about liberty and the like. One thing that has come up in recent biographies is the fact that Voltaire, along with his talents in philosophy and literature, was a sharp businessman and quick to take advantage of a good deal. At one point he executed the eighteenth-century equivalent of insider trading and garnered a large profit. In one of his business dealings he was outsmarted by sharp Jewish moneylenders; and although that probably had little to do with his basic antisemitism, he never did like to lose. Furthermore, recent critics have accused him of racism. And, indeed, among one of his business deals, he invested in a share of a slave ship.

Not long after our earlier meeting, I asked my Dr. Pangloss about Voltaire's antisemitism.

"Sure," he said. "He was also an apostate; he despised the Church, reviled priests and bishops, slandered the pope, and happily pointed out the absurdity of all religions. And anyway Voltaire lived in Europe in the eighteenth century, and western culture has had a problem with Jews ever since the reigns of Augustus and Hadrian, although one has to wonder why a liberal-minded historical figure like Voltaire would publicly announce his views."

"On the other hand," he continued, "look at Voltaire's treatment of the Lisbon earthquake. All the Christians in church and all the reprobates out on the streets after a drunken Saturday night, also the Jews; it's Sunday morning, and at just the right time all the churches collapse, killing the good Christians, and leaving the Jews and the reprobates alive. If God is kind, what caused *that* to happen?

"Voltaire put a good twist on that one," he told me. "'The Jews did it; we must hold an auto da fé.' But actually Voltaire made that up. The Church fathers didn't blame the Jews. They claimed it was God's will, and who are we to challenge the Lord?"

In *Candide*, Voltaire mocked that explanation, too. The poor Christian proselytizer, James the Anabaptist, who is traveling with Candide at the time, falls overboard and drowns in the rough seas during the earthquake, while the irreligious Candide survives. Dr. Pangloss tells Candide that their friend James's death doesn't matter because it was an act of God, and God is good, so it must be okay. Furthermore, he claims that God created the harbor of Lisbon for James to drown in. "This is the best of all possible worlds," he exclaims, as usual. According to Dr. Pangloss Rosen, "Voltaire probably invented the auto da fé scene to point out the absurdity of organized religion."

Our discussion of Voltaire's antisemitism launched my Pangloss into one of his wide overviews of social history.

"You've got to be careful with the personal lives of writers and artists, or you'll have to ban half the works of western culture. Think of the many drunks, wifebeaters, and antisemites among the artists and writers—even murderers: Caravaggio, for one, and Christopher Marlowe. And how about Shakespeare? We'll have to ban *The Merchant of Venice*. What about Bach's *Passion according to Saint John*, with its antisemitic overtones (or maybe undertones)? What about Wagner? I mean the list goes on—the Dreyfus affair; T. S. Eliot, antisemite; Ezra Pound, antisemite. What about the French novelist Céline's antisemitism? For that matter, what about the whole nation of Germany in the late 1930s? And then look at Voltaire and *Candide* and the absurdity of holding autos da fé and blaming the Jews for the earthquake (and even mistaking Dr. Pangloss for a Jew and hanging him)."

It was a good speech in spite of its reactionary undertones—typical of my Dr. Pangloss.

Voltaire's idea of cultivating a garden in the midst of hard times was not limited to *Candide*, as I mentioned earlier. Voltaire used gardens as a metaphor or emblem for local altruistic reform, one that stood in opposition to the established order of the Church and blind conservatism. In his garden he was not retreating from the horrors of reality; he was making a statement of resistance.

I have always felt the same. Nonetheless, compared to the world beyond the garden walls, I and the people in my demesne live in a park. Puttering around on warm mornings, I sometimes imagine myself as an old English gardener in baggy corduroys, a worn-out waistcoat, and a cloth cap. Or sometimes I picture myself in rumpled suit pants and a beret, a reflection of a gardener I used to see almost every day when I worked on Corsica. He had an immaculate vegetable plot near the grounds of an old hotel, and he must have been a very early riser because I would see him hoeing his beds when I went to the town boulangerie to pick up bread for the day. In my memory neither of these models worked very fast, nor do I. What's the rush, if the world is coming to an end? I might as well slow down and enjoy it while it lasts. Unfortunately, this is the position of any number of people, both young and old. "Not our problem" . . . until it is.

In 2022, the world disasters continued. Huge sections of the West, the Midwest, and the Southwest baked in drought and record heat waves. Then the rains came, bringing with them five different 1,000-year floods, the result of a phenomenon known as weather whiplash. In such conditions, prolonged hot air bakes the soil into hardpan. Then, when the rains finally do come, the waters cannot sink into the earth and become floods.

The same thing happened in Pakistan, which, along with Bangladesh, always seems to suffer from particularly brutal environmental disasters. After a record-breaking drought, the country was stricken with a mega-monsoon, also record-breaking, with far heavier rains than usual. Weather whiplash triggered the worst floods in Pakistan's history, leaving a third of the country underwater, farms and villages flooded, and 30 million people uprooted and homeless.

Problems continued in the United States, too. In early autumn, an immense destructive hurricane hit southwestern Florida, not record-breaking (for once) yet strong enough for climatologists to agree that powerful storms of this sort were the new normal.

The drought that plagued Sunny Bank Garden in the late summer of 2022 was one of these projected climatic events. After the rains and the floods of spring, the soil slowly began to dry. Despite an absence of rain, the garden carried on. Late spring was warmer than usual, and then we had two minor heat waves—nothing compared to the Midwest's, but still we had no significant rain to cool things off. By early August, the rains had quit altogether. The town imposed restrictions: no watering between seven in the morning and seven in the evening. The sun continued to blast, the earth dried hard, and the town expanded its restrictions. No lawn watering, then no sprinklers at any time; hand watering only. Withered suburban lawns became a symbol that homeowners were accepting the seriousness of climate change. Green lawns were clear evidence that cheaters were watering in spite of the regulations.

I have a private well—no town water–and I never water the lawn anyway, and it's not much of a real lawn. I encourage local groundcovers such as creeping thyme, ajuga, and certain species of so-called weeds, including a plantain that I used to dig out until I heard a lecture on local butterflies and moths and learned that a rare species of moth feeds on that sole species of plantain.

My garden finally began to react to the drought. I have a beautiful weeping beech overhanging the drive up to the house. Its upper branches began to die, always a sign that a tree is in serious danger. One of two katsura trees on the south side of the drive shed its leaves entirely; the other was clearly stressed. A big kusa dogwood at the head of the drive never flowered. The leaves of a prized pink dogwood curled up, as did the leaves of a unique pink-flowering hawthorn. One of my Carolina snowbells lost all of its leaves by late August.

As the drought began to take its toll, I realized that the plants that were most affected, most at risk of dying, were the shrubs I had planted in early spring, including some of my roses.

Still, I continued my rosebush memorials. After one disaster— another huge fire in New Mexico—I bought a Princess de Monaco rose and planted it in a new bed bordered by pear trees. In spite of the fact that I watered it regularly, as with the roses I had planted earlier in the drought, it, too, began to die, so I moved it to a nearby rose bed. It struggled along but never flowered. Some of the other roses I had planted that spring had flowered profusely in the earlier rains but failed to grow at all after the drought hit—there was no second blooming.

The aforementioned animal tombstone is located in a shaded area backed by old oaks. To create this section of the garden, earlier in the spring I had cleared a short path through an unmanaged tangle of ferns and brush and laid out an allée of rhododendrons and andromedas and transplanted some of my existing shrubs into this spot. I also planted four young cryptomeria—Japanese temple trees—two on either side of the boulder. Cryptomeria are very slow growing; so if they survive, the only ones in my cohort who will get to enjoy them will be my grandchildren—provided they stay put. If not, I can only hope that the new owners of the house will love trees and put up with the eccentricities of this garden.

This shaded garden lies at a distance from the house. At first I hauled out buckets regularly to water the plants, but eventually I decided to let the weather do the watering. Under typical conditions that would have sufficed, but not during the drought. So I transplanted all of the young bushes into pots and moved them to a bed at the head of the drive, where I could reach them with a hose. I did the same with other flowers and shrubs I had planted that year and managed to keep them green until late September, when the rains finally arrived.

I should say that this might be the worst possible land to plant what one early nineteenth-century writer termed a "pleasure ground"—that is, land laid out for enjoyment rather than profit or household use, as with a farm or a kitchen garden. That's because the property is located at the top of a drumlin, an egg-shaped hillock underlain by clay, sand, and gravel.

Drumlins were formed about 15,000 years ago. As the last glacier retreated, it left behind pockets of gravel, sand, and rocks that were plowed up into mounds that mirror the path of the receding ice sheet. Drumlins are common in the region west of Boston and also in Boston Harbor. The islands of the harbor are actually the tops of drumlins.

According to geologists, the climatic forces that brought on the last glacier about 60,000 year ago fit into the normal pattern of climate change in Earth's history: periods of wet, warm conditions, droughts, long ice ages, tropical eras, and the like, all of them natural and sometimes including massive extinctions. In our time, however, the geological and climate change forces that have shaped Earth's history are now measured in decades, and the most rapid of these changes have taken place in the past thirty years.

The best evidence of these changes (as if we need any more evidence) lies in recent findings by the World Meteorological Organization, which has determined that sea-level rise in the coming few centuries is guaranteed, even if the nations of the world agree to stop releasing greenhouse gases immediately. Sea levels have been rising slowly since the late nineteenth century, but the past thirty or so years have seen the most dramatic increases; and according to studies, they are still accelerating. Paleoclimatologists—those who study the history of ancient climate conditions—have determined that the Earth is exhibiting its highest average temperatures in recorded history. The planet is now warmer than at any other time in the past 125,000 years.

Back in 1955, engineers developed the first computer-generated climate models. Using a method they called feedback loops, they were able to input certain established facts such as the rise of carbon dioxide in the atmosphere since 1850. By coupling those data with information about the subsequent rise in worldwide temperatures, the loss of sea ice, the worldwide meltdown of glaciers, and the rise of sea levels, they could generally predict what the future would hold if the pattern were to continue.

Twenty years later these models were refined, and they continue to improve year by year.

I knew very little about feedback loops in the 1980s, when I started researching stories on what was then called the greenhouse effect. I only began to understand how they worked after attending a conference on climate change at Clark University. The main speaker on the subject was John Holdren, who later became the Obama administration's senior science advisor. During his talk, Holdren's powerful computer projected its findings up onto a large screen, which made it easy for people like me to follow what he was doing. Moreover, unlike other speakers in his field, he spoke comprehensible English. Technical language notwithstanding, I could understand what he was talking about, and it was scary.

What I remember best, but found hard to believe, was what the world would be like roughly twenty-five years out. The computer models predicted record-breaking heat waves and droughts, massive floods, uprooted populations, sea-level rise, and warming oceans—in short, everything we are now experiencing, only maybe even a little worse. Climatologists were not aware back then that the Arctic would be melting four times faster than they had predicted.

I should add that there was some hope as far as the timelines were concerned. At another climate lecture I attended, this time at the Woods Hole Institute in the early 2000s, the speaker and his computer calculated the rate of ice loss on Greenland's ice shelf. I've forgotten the researcher's name, but he (or at least his computer) predicted that the ice sheet might vanish within seven years.

The facts of his findings still hold true: the Greenland ice shelf is indeed melting at alarming rates. But his timing was wrong. Now, twenty years later, most of the shelf is still intact, and the newest studies place the tipping point sometime in the next few decades.

The loss of this ice will wreak devastation in northern Europe. Part of the problem is that fresh water floats above salt water, so the input of fresh water from the melting glacier will create a barrier that will block the natural heat-exchange cycle of seawater, a phenomenon that keeps England and northern France warmer than land masses at the same latitude in North America.

Furthermore, there is alas alarming news from studies tracking what is known as the Atlantic meridional overturning circulation, which works like a huge underwater conveyor belt, carrying warm waters up to the North Atlantic from the tropics and sending colder water southward along the ocean floor. Some researchers say that the input of cold fresh water from the Greenland ice sheet and the Arctic Sea could abruptly shut down the system. When this will happen is the big question.

Research carried out by the UN Intergovernmental Panel on Climate Change suggests that the system will not fully melt until the end of the century, although the recent tipping point could come much earlier. Newer studies suggest it could occur as soon as 2050. Whenever it happens—which all the studies agree it will—the climate of Britain and northern France will match Canada's, which is to say that England will resemble Newfoundland. This would mean, among other unfortunate results, crop failures in the United Kingdom, major inundation of the East Coast of the United States, and an alteration of the wet and dry seasons in the Amazon. It would include a farewell to the great gardens of England, not to mention the loss of vineyards in northern France as well as fruit crops such as peaches, figs, and plums.

What can one possibly do with all of this gloomy evidence?

I went out and bought a new rose, a recently cultivated variety called Sunblaze, a miniature pink rose that I planted in a flowerbed just outside the kitchen window.

One of the things I came to appreciate about the little grove abutting the northeastern side of my territory was the fact that I could go there and do nothing. I didn't have to work to maintain it, other than clear away a fallen branch or two from time to time. I could rest there, alone. It was a good place to escape; and because I used to go there on a regular basis, over time I began to think of this hidden grove as my own property.

Thoreau had the same inclinations. He would make a wide daily circuit of the fields and forests around Concord, later writing that he had "owned" every farm in Concord, or at least the ones he'd taken a fancy to. He took possession of them only in his mind, but that did not keep him from enjoying them; he was an imaginative trespasser.

I suppose I experienced the same feeling. I think I may have been the only one who actually knew of the existence of this little place apart. Over time, in my mind, I effectively expanded the boundaries of my garden and began to think of the land, which I would walk over almost every day, as my demesne. Although the legal master of the estate was the kindly steward, Lord Findlay, I imagined myself to be a rebel peasant, like Robin Hood—a legal resident of the common forest. The other houses, gardens, and forest tracts were merely part of my holdings, set within the larger estate of Scratch Flat, one of four divisions of the town, along with Hog End, the East End, and the town center.

As I've said, back in the mid-seventeenth century, this area was part of the Nashobah Plantation, a tract of land held by a group of Christianized Indians. The territory was originally controlled by the powerful chief Nanepashamet of the Pawtucket Confederation. After he was killed in a conflict with the violent Tarrantine people of northern Maine, his wife, whom the English called Squaw Sachem, took over his lands. In 1635, she sold a section of her holdings to the English, and the land was incorporated as the town of Concord, the first inland settlement of the Massachusetts Bay Colony. Scratch Flat was part of that holding.

Today, nine people live in the five houses within my demesne. Along with three dogs and two housecats, the land has a large population of transient and resident wild animals, including newcomers that have moved in over the past few decades such as bears, otters, fishers, and coyotes as well as several species of birds that have spread northward with global warming. Also living in this territory during the past 15,000 years were three different cultural groups of Native Americans. One could argue that this land technically should belong to the Native people (and also to the animals, according to a legal argument promoted by the progressive Supreme Court judge William O. Douglas and a few contemporary animal-rights attorneys). But under the American legal designation known as *fee simple*, the land is now "owned"; it is private property, although nearly 150 years had to pass before it acquired that label. Before that it was, according to English law, common land.

In the 1640s, the English took over the land of the Massachusetts and Nipmuck people who were already living here. English law decreed that the vast untrammeled forests, which in that era stretched from the Atlantic coast westward into unexplored sections of the interior, belonged to the king, represented by the Massachusetts Bay Colony. However, according to a detail of English common law known as *alollodial rights*, the cultivated lands that the Natives were farming belonged to them and therefore had to be purchased from them. (Never mind that the English did not pay the Natives as they would have paid one another. Instead, they used a few beads, trinkets, and hatchets to gain sizable tracts of land.) At the same time the English did not quite understand the fact that the Native people had so little understanding of the land's monetary value that they saw nothing wrong with selling it more than once.

This history may seem superfluous to the current status of my garden and my theoretical demesne, except that these new land designations established the legal background for things to come.

I was away in March of 2022 and isolated from the news, but when I got back in April, I learned that it was bad: a series of political conflicts, including a war in Ukraine, increased numbers of refugees fleeing environmental and political issues, assassinations, contested elections, and, in the United States, a sort of internal and totally pointless anarchistic civil war in which people armed with military weapons made raids on churches, synagogues, gay bars, and elementary schools, killing as many children and adults as they could. It reminded me of the world that Candide and Dr. Pangloss traveled through before they got to the garden in Turkey.

Things were no better on the climate front. Along with the continuing massive fire in New Mexico, there were fires in Texas and southern Colorado. There were the record-breaking early heat waves in India and Pakistan. Day after day the temperature in those countries rose above 100 degrees Fahrenheit and did not drop below 95 or so at night. These heat waves had melted glaciers in the Himalayas, leading to downstream flooding that was stunting the wheat crop, further increasing food shortages, and causing widespread power outages.

Thoreau says somewhere that he felt lucky to have been born in Concord and "in the nick of time." The same, I suppose, has been true for me. The wheel of fortune landed me in a green sanctuary in a region that is experiencing fewer climate disasters than other parts of the world, and on the day I took in all of this bad news, the Vicarage Garden had never felt so welcoming. It flourished on its own even without my constant care, a true sanctuary.

Climate scientists of course blamed the unusual increase in heat waves on global warming and were again predicting even worse conditions in the coming decades. UN climate studies had determined that, in spite of the stated goals of the Paris Accords, fossil-fuel emissions had increased that year. Nations that had pledged to reduce carbon emissions instead broke a record for carbon-dioxide release, which by that time stood at 417 parts per million. Another study indicated that unless alternative energy sources could be put into operation in the next nine years, global warming would continue, passing a critical warming threshold and triggering even more catastrophic events.

Some would argue that all of this math and theory is of little concern in daily life. And it is true that, as the world heats up, the days and weeks roll on and the pattern of daily life disguises the realities of the coming deluge. It's all far away, some claim; it's all just guessing and unreliable science. Doomsday predictions come and go, but life goes on.

And yet.

The day after I got back from my trip, I went out on my usual circuit of my demesne. I started on my own legally held land, wove through a dark grove of hemlock and woodland trails on the northwestern side of my property, turned eastward, and crossed the road. But before I reached the trailhead I saw a large sign announcing proudly that my personal sanctuary, a cherished, even sacred, section of my demesne, was up for sale.

PART TWO

Beyond the Garden Walls

CHAPTER NINE

THE BETTY PRIOR ROSE

As if to celebrate in a perverse way the proposed destruction of my local woodland, a devastating wildfire in northern California burned more than 2,000 acres of forest, and an uncanny heat wave struck Canada and the Pacific Northwest. Record-breaking high temperatures were recorded across the region. In any normal year, before industrialization and the rise of global warming, heat waves of this sort would be cast as a weather event, not unlike a serious storm, a heavy winter snow, or a dry summer. Droughts, or even decade-long periods of cold winters, such as those that occurred in Europe during the Little Ice Age between 1300 and 1850, once occurred naturally. But the uncommon climatic events of that period were generally regional, not worldwide.

Nowadays, current prolonged record-setting weather events are global and are caused by an increase in the Earth's temperatures. The most obvious examples are the slow retreat of glaciers as well as melting ice in the poles, on Greenland's ice shelf (Earth's largest glacial ice sheet), and on the massive Thwaite's ice sheet in Antarctica—the so-called Doomsday Glacier that has experienced a net loss of more than 600 billion tons of ice in recent years.

❧

Now that my neighboring sanctuary, the enchanted forest, was threatened, I made a point of visiting the woods every day. In late spring, one of my birdwatching friends told me that he had heard, even seen a few times, a blue-gray gnatcatcher along the western edge of the brook. So early one morning, I walked over to the forest to spend an hour or so listening for the bird's whiney, whisper-like call, hoping to spot one.

Blue-gray gnatcatchers used to be strictly a southern species: I would often see them in the woodlots on one of my family's farms on the Eastern Shore of Maryland. But like so many other southern species, climate change has allowed them to expand their range northward during the past few decades. I had seen them in woodlands south of me, near the Sudbury River, one of Thoreau's haunts. Now they had apparently moved into my territory.

After wandering around a bit, looking into the treetops and listening for birdcalls, I spent the rest of my sojourn mostly sitting quietly on a large glacial erratic boulder in the southwestern part of the woods, away from the brook. I saw a number of local birds there: a catbird, a towhee, a few buzzing little warblers that I couldn't see well enough to identify, a yellow warbler, and a Maryland yellowthroat, along with the usual woodpeckers, blue jays, finches, and other common species. But there were no blue-gray gnatcatchers. I decided to give up and go home for a late breakfast.

Just as I was exiting the woods and stepping out into the field, I saw a large Mercedes SUV pass slowly by, stop, and back up. A blond heavyset man got out of the car and crossed the field toward me with a determined stride, his head lowered like a bull.

I waited.

His opening salvo made things all too clear. "Private property!" he said loudly.

He wanted to know what the hell I was doing there in the woods.

As a seasoned trespasser, I have learned to read the character of those who feel it necessary, for one reason or another, to inform me that their land is private. I could see where this man was going; and to defuse him, I played the part of a slow and innocent birdwatcher.

"I was looking for blue-gray gnatcatchers," I said. "They are beautiful little birds, very rare. Do you know them?"

He didn't answer.

"That's private property, okay?" he said. "These woods. You can't go in there."

"Private property?" I asked, taken aback.

"Yeah, private. Keep out."

"But these are Charlie's fields. He used to graze his heifers here. One of them escaped a few years ago."

"Well, whoever Charlie is, he don't own that land."

"Terribly sorry, I thought this land was Charlie's, the old farmer," I said. "You see I walk here daily on my circuit, looking for birds. There are some beautiful and rare species here, including reports of the blue-gray gnatcatcher." (I was playing the nerdy birdwatcher angle to the hilt at this point.)

"Yeah, well don't go in there, and get off this land now. You live around here?"

"Yes," I said. "I am the groundskeeper for Sunny Bank Garden, a sanctuary for birds. We get some very interesting species there. Also rare crickets and moths. I make a circuit through these lands, they're all a part of Scratch Flat, and I am researching the various species of birds and mammals that occur here. I could show you around if you'd like."

By this time, the man, whoever he was, had calculated that he was dealing with someone who was slightly off-center, possibly feeble-minded, and he grew kinder.

"You're not supposed to go in those woods anymore," he said.

"But why not? I've been walking there for twenty years, searching for rare species. I saw a hooded warbler in there last year. Also, a prothonotary warbler! Very rare."

"I'm sorry," he said. "But you can't go there anymore. That's my property now."

"You own it?"

"Yes, but I'm selling it."

"But Charlie likes to have me walk there, and he appreciates the fact that that I'm keeping a watch on his land."

"That right?" the developer said. He was getting tired of all this.

"Well, yes," I said. "Charlie doesn't get out here as often as he used to. He's older, don't you know?"

I didn't mention that this Charlie, the original owner, had died three years earlier.

"Listen, pal, Charlie don't own this land anymore. I do, so don't go in there, all right? Please don't go there."

"Don't go in the woods?" I asked haltingly, as if he had said "stop breathing."

"Not those woods. Private property."

I wanted to tell him that the Native American people who had lived on this land for thousands of years did not believe that one could actually own land, but I decided to give up. I looked back at the woods, then looked at him, and shrugged, dumbfounded.

"Stay out," he said with finality.

He turned and marched back to his car.

I was tempted to tell him that he had not actually seen me in the woods and therefore could not legally charge me with trespassing. But because he seemed to indeed be the owner and developer, I held my fire. There would surely be more public confrontations to come.

Of course, threats notwithstanding, I went back to the woods the next day, and the day after. But when I went back the following week, the property was surrounded by "No Trespassing" signs. And thus the woods became the Forbidden Forest, which, again, made the place all the more enticing.

Not long before my encounter with the developer, I was leading a group of international students from a Harvard graduate-school class through the Estabrook Woods in Concord. As we crossed an open clearing, an old man stepped out of the brush and glared at us. I knew this man. He was the patriarch of an old Concord family who had opened his property and other tracts of land to the public. He used to make a daily woodland circuit of his former holdings, perhaps reviewing his now un-privatized lands, hobbling along with a cane, his glinty eyes watching the ground for roots and rocks.

I explained that I was taking the group of foreign visitors through the woods, identifying local plant and bird species and detailing some of the American legal means of protecting wild lands.

He nodded in approval; but as we bid him goodbye and moved on, the old prophet raised his cane. "Ye cannot own land," he called after us.

(Actually, he said *you* cannot own land, but *ye* is more descriptive of the encounter.)

Many cultures around the world, including the Native Americans who once settled that very spot, would agree with him. But how did land come to be owned, as one owns a car, for example? I learned this through personal experience long before I was able to read complicated legal documents.

I grew up in an old Hudson River town not far from New York City, which, in the late nineteenth century, was a ferryboat suburb for Wall Street magnates. The houses here were high and gabled, the trees were higher than the houses, and the lawns were deep green, the flowerbeds overgrown, the fishponds weedy and populated by frogs that ducked beneath the murky waters before you could catch them.

The town had once been moneyed, but after the Depression the gardens went into decline, as did some of the old carriage houses and barns, not to mention the people who inhabited them. The old families still lived there, surviving on dwindling trust funds. When I was growing up there were holes in the privet hedges, empty woodlots, foundations, and, most intriguingly, ruined estates with crumbling pillars, marble steps leading to weed patches, and broken greenhouses stretched along the high cliffs of the Palisades.

Like most suburban and exurban children, my friends and I were turned out to roam shortly after breakfast. I was not expected to return until dinnertime but would scrounge lunch at one household or another. It was here, in this undeclared version of common land, that I undertook my first resistance to the destruction of woodlands.

On these free days, my friends and I used to collect together and set out to seek adventure. Usually we ended up in an unhoused woodland consisting of ancient beeches, sweetgum, sourwood, oak, and maple, where hawks nested and foxes had their dens. There were no supervising adults in this forest. It had been a haunt of my older brothers and their band for years and probably a hideout for kids before that. It was private property, but in the free-range territory of disinterested householders of the town, we were not aware of any legal designations. Parlin's Wood, as far as anyone knew, was common land.

But one Saturday morning, reality intruded. We arrived at the forest edge only to be confronted by "No Trespassing" signs and a huge billboard announcing proudly that our version of Robin Hood's Sherwood Forest was to be the future home of "Parlin's Hill Estates."

Young eco-terrorists that we were, we tore some of the signs down.

To make a long and complicated story short, our act of eco-terrorism ultimately alerted the adult world to the fact that this unspoiled twenty-plus-acre forest was to be leveled. And so began a long battle to save the woods.

That day, after our early version of environmental resistance, we were captured by the sheriff's men and turned over to our parents for punishment. But among the other adults who lived near the woods there was a well-known nature photographer named Campbell Norsgaard who was associated with *National Geographic* magazine. He knew of our sojourns in the woods. In fact, he had at one point shown me a hawk's nest and invited me to ascend a long ladder he had built to photograph the fledglings. I daresay he approved of our trespasses, and in time he took up our defense, and the defense of the woods. Other adults joined him. Lawyers were hired, and eventually the woodland was saved and, some years later, turned into a nature center.

I did not know back then that, by inciting this revolt against the doctrine of private ownership of land, my friends and I were reenacting ancient conflicts. But given space and time and a lack of adult supervision, children will replay history. The features of the legal case that would come to pass—the fight to save common land and the clash between excessive private enterprise and the commonweal—have been part of history ever since the rise of the first human settlements and the creation of a ruling class.

Written laws governing the use of land go back to the Code of Hammurabi. They were refined by the Roman emperor Justinian, English common law, and the forest laws of William the Conqueror, then further refined by the Napoleonic Code and finally, and dramatically, by the American Constitution. Most of these restrictions led to uprisings among those affected by the changes, usually the poor and disenfranchised or the native people of regions overtaken by foreign imperialists. As far as their influence on American law is concerned, perhaps the most significant were the revolts against the forest laws instituted in 1066 by King William that forbade the peasants from using the "vert and venison," forest resources that had till then been crucial to their livelihood.

Several Anglo-Saxon rebellions took place shortly after 1066, and not all were related to land use. But one of the demands of Watt Tyler's famous peasant uprising in 1381 was a revocation of the forest laws. The Jacquerie, a French peasant rebellion in 1358, was also a reaction to land controls.

But as far as my story is concerned, the best-known tale of revolt is the legend of my childhood hero, Robin Hood, who was what we now might term a radical environmentalist. His dispute with the king developed as a result of changes in land-use laws in England that began in the eleventh century. During the period of the most popular Robin Hood legend—the beginning of the thirteenth century—King John owned more than Sherwood Forest. He owned all of the lands of England, as well as the fish in the streams, the rabbits in the greenwood, and the deer and wild boar of the forests. A century and a half after William the Conqueror's reign, Robin and the other Anglo-Saxon yeomen and barons of the northern districts were still reacting to Norman rule and the draconian forest laws.

By the thirteenth century, the English monarchy had introduced new policies that permitted the transfer of common forest lands to wool producers and treed common lands to private timber operations, further curtailing the rights of the commoners. So, according to early versions of his legend, Robin Hood—always a forest dweller—sallies forth to resist the king and succor his fellow yeomen. Not only does he poach deer and live a free life in the wild, but he leads a revolt that restores the land to the peasants. Never mind historical fact, it's a good story; and as the well-known Robin Hood scholar Stephen Knight points out, true or false, it accurately reflects the spirit of the age.

With the signing of the Magna Carta in 1215, feudal control of lands began to decline, and what some consider an even worse condition began to evolve: a legal system known as *socage*, in which tenants could pay for land use in rents rather than in service or in kind as it had been under feudalism. Throughout Europe and North America, egalitarian, primordial land-use systems were supplanted by legal doctrines that favored a few wealthy patrons. As a result of the acts of enclosure, a series of laws that Parliament began instituting in about 1600, English freeholders were evicted from the common lands that had existed even under the constraints of feudalism. In North America, the Native tribes' ecologically sound method of divvying up resources according to supply, was displaced by European systems; and in Latin America, the feudalistic Spanish *encomienda* not only replaced more communal Native agricultural land traditions but killed or enslaved tribal members in the process. Eventually, a theory of private ownership of property developed in the newly created United States. The policy of fee simple flourished and was forged into law in the Fifth Amendment of the U.S. Constitution.

The land-use laws may have varied from place to place, but the endings of the stories are the same: the common folk were denied the use of common land.

In reaction to all these changes, a long line of Robin Hood–like activists, rebels, well-intentioned bandits, and latter-day environmental guerillas stepped up to defend the rights of the poor and the conservation of open land. This group includes the contemporary environmental activists of our time, who commit crimes in defense of the environment. All of these players, now and in the past, believe that land should be open to all the people, not a select few.

The fight for the ancient tradition of common land has not ended. But along with the preservation of wildlife and other benefits, there is now one more reason to save as much land as possible: because of the role of trees in the face of climate change.

Climatologists believe that young growing trees and especially old-growth forests are the best and cheapest counter to increased levels of carbon dioxide. Thus, reforestation projects are under way in many, but not all, nations connected to the Paris Accords.

A single mature tree will consume more than forty-eight pounds of carbon dioxide in a single year. I've got more than three hundred trees in my garden, which means that this single acre-and-a-half tract is absorbing about 14,400 pounds a year. I think the little nearby woodland sanctuary consists of about twenty acres of trees at about three hundred trees per acre, which means it is absorbing about 691,200 pounds a year—not much in comparison to the 34.81 billion metric tons of carbon dioxide that are released worldwide each year, but every little bit helps, as they say.

Put another way, if developers were to cut down the trees in the local woodland, they would be adding 691,200 pounds of carbon to the atmosphere. If they were to replace the trees with twenty suburban houses, they would be adding about 7.5 tons a year—and this figure doesn't count the carbon releases created by construction.

Not a good prospect as far as countering climate change is concerned.

In this town, and in many of the surrounding towns, developers rarely propose to replace a tract of woodland or farmland with a housing development without meeting resistance. One group or another will rise up in defense of nature. Most of these legal disputes are lost, but not all.

In this town, the local land trust usually leads the defense, but in difficult battles, larger statewide or occasionally national land-preservation organizations have joined in.

And so it was with the announcement that the sylvan grove where the blue-gray gnatcatcher nested was about to fall to the bulldozer: a campaign to save the woods was organized.

CHAPTER TEN

FAREWELL TO A WILD GARDEN

O ver the course of the summer, I continued to trespass in the woodland; but one morning, as I turned into the fields, I heard a backhoe working in one of the older pastures, about fifty yards southeast of the little glade. I decided to turn around and instead enter the wood by bushwhacking through a tangled thicket owned by the local sportsman's club. Later that day, I went to our town hall to find out what was happening with the land but was told that the tract in question was actually located in a neighboring town to the north. So I went to that town hall, found the proper office, and was informed that not only was the grove threatened but the whole former farm was scheduled to be developed into a typical boring suburban development of fifty new houses. The Forbidden Forest was only the southernmost section of the land.

When I moved to the town, this was a farm called Sunny Bank, owned by the aforesaid Charlie mentioned during my encounter with the developer. At that time it consisted of a hayfield and a pasture for heifers. There was an old barn with a caved-in roof, and there were a number of old chicken coops and other outbuildings scattered around the property, which dropped down to the marshes of Beaver Brook in a series of natural terraces.

After Charlie died, the fields were deserted. The open land began to grow up to brush after a couple of years, creating a particularly rich ecosystem known to ecologists as an early seral stage. Old fields of this sort consist of grasses, forbs, shrubs, and a few sun-loving trees such as gray dogwood, alder, and black cherry, and they are among the most diverse of ecosystems, providing habitat for a high number of birds, mammals, reptiles, amphibians, and insect species.

Barn swallows nested in the old barn, and raccoons lived in the old chicken coops. Foxes and groundhogs were common, as were short-tailed weasels, meadow voles, white-footed mice, and moles. A black snake lived in one of the old foundations, and garter snakes were common, as were brown snakes, ring-necked snakes, milk snakes, and rare (at least in these parts) green snakes. Pickerel or leopard frogs (I could never tell which) appeared in long grasses in late summer, along with a host of night-singing insects, including snowy tree crickets, bush katydids, and field crickets. There were monarch butterflies in the milkweed patches, and one summer day I found a buck moth there, an endangered species in this area.

A shallow pond on the lowest terrace was populated in spring by a host of spring peepers, spotted salamanders, and fairy shrimp. And on March evenings, woodcocks favored the area for their mating dances. They would circle on the cleared ground, strutting and bowing and chirping in a nasal beep, and then take flight. You could see them against the sky at dusk, and then watch them descend in a series of loops, their wings whistling in a fluted chirp, before repeating the dance.

Prairie warblers and golden-winged warblers and bluebirds nested there, song sparrows and white-throated sparrows were common, and in later summer flocks of Savannah sparrows would whisk over the land. On any given day during the growing season, cardinals, chickadees, robins, goldfinches, house finches, chipping sparrows, song sparrows, field sparrows, cedar waxwings, Carolina wrens, and even increasingly rare meadowlarks and kestrels could be seen flying or nesting or feeding in the old fields.

It was a known fact (to me, at least) that the famous English pirate William Fly, who used to raid vessels in the shipping lanes off the New England coast, had buried a chest of Spanish doubloons somewhere on that land, and I and my young (and innocent) archeological assistants used to search through the old foundations and cellar holes for the treasure. My oldest brother, James, used to set up photo shoots on the property in which local children would replay scenes of history, including captures and daring rescues, escapes, and raids involving William Fly.

There was a dry and deep foundation with a south-facing wall, and on sunny days in early spring we could collect there and bask in the warming sun, away from the wind, along with the black snake and a milk snake, who would also emerge to take the sun.

For its part, the Forbidden Forest was an alternative and unique forest ecosystem. To say that the woodland tract contained many species of trees, shrubs, and wildflowers may be a meaningful description of the woods, but it hardly conveys the rich biodiversity of the place. More engaging would be an accounting of what would be lost to development:

To cite one example, the trail through the woods led through a veritable forest of ferns. Here were lacy woodferns and the sensitive ferns, first to die back in autumn; also the delicate lady fern, New York fern, stately royal ferns, and the clinging rock polypody, which stays green all winter. I have found ebony spleenwort and the ostrich fern with its feather fronds; and just below the old trunk of a fallen white pine there were more species of wood ferns, as well as Christmas fern, which also stays green in winter, and interrupted fern, and hay-scented fern in the sunny clearings.

At the edges of the grove, where the aspen leaves fluttered in the breezes and the red maples burned in bright red flames in

April, I could find meadowsweet and steeplebush, elderberry, nannyberry, blueberry, cranberry bush, arrowwood and sweetfern. Also four species of oak trees, three species of maples, and hickories, slippery elm, sprouts of chestnut trees, ash trees, alders, white pine, hemlock, and spruce. All this and more was scheduled for execution.

I wandered around the woods that month nostalgically contemplating the doomed land—the old oaks and hickories and maples soon to be felled, the rich topsoil stripped, along with the resident animals, killed or evicted to new territories, where their chances of survival were slim.

Knowing what was coming, I began a dedicated search for rare and endangered species in hopes of finding something, anything, that could stop or at least cut back on the size of the development.

I was not the first to raise the alarm over the pending sale. The local land trust and other conservation-minded individuals in the community spread the word and began making plans to try to save the woods. But the land in question was privately held—and therein lay the problem.

I won't say that I wept and moaned on that late spring morning after I learned that the forest and old fields were to be developed. What I felt was a sinking sense of fatigue. Here we go again, I thought: yet another battle to save a plot of wild land.

I can't count the number of environmental and land-preservation legal battles I've been involved in. It seems to have been a family tradition, actually. My mother got me interested in birds and flowers, and my father, who once owned a large farm on the Eastern Shore of Maryland, was peripherally involved in land issues. My middle brother took up the environmental flag and engaged in uncountable battles to save land in the Rochester, New York, area. He and his group even took a case to the

Supreme Court, although he lost and the federal government rammed a highway through an Olmstead-designed urban park. (Not incidentally, perhaps, this park was regularly used by the residents of a nearby African American neighborhood.)

In his way, my oldest brother, James, was the equivalent of an early environmental terrorist. He used to pull out survey stakes from soon-to-be-developed plots and talked often of dumping sugar into the gas tanks of bulldozers to clog engines and halt development. He also used to publish environmental cartoons that depicted bomb-carrying Canada geese attacking wetland developers or the uprisings of offended muskrats and beavers. At one point, he created a fake roadblock to keep tourist buses from passing through the quiet neighborhoods of his town to get to a popular lobster restaurant, thereby raising a protest in the village that eventually got the buses banned, and when I was twelve, I published a popular letter in the local newspaper lamenting the loss of an ancient copper beech, chopped down to make way for a new furniture store.

The days and nights are long and all-consuming during these periods, and all I wanted to do in these last decades or so before the world collapses was to stay home and cultivate my garden. But our times do not grant such opportunities, and so—off to the front.

I began thinking about the plants and local amphibians that had been there in the past. Knowing that one or two might be rare or endangered, I decided I should make—as best I could by myself—a biological survey.

I'm not an ecologist and not even that sharp a naturalist, but I know what I'm looking at and can spot something I've never seen before growing among the common species of the demesne. Perhaps I could find a rare salamander, or a rare butterfly, or a dragonfly, or perhaps an obscure mushroom, as yet undiscovered. These things had been found locally during other land-based legal battles and had helped to save acreage. One friend had discovered a rare dragonfly during a legal case involving wetlands. In another interesting case involving plans for a well-funded shopping-mall development, activists, after a drawn-out legal battle, saved a vernal pool where a threatened population of wood frogs was breeding. (This turned out to be a Pyrrhic victory because wood frogs were reluctant to cross parking lots or even sheared lawns, for that matter, and the colony died out.)

Just downstream from the brook below my neighboring woodland, where the stream enters a lake, herpetologists had found an endangered Blanding's turtle. But it would be hard to make a case that the vernal pools, springs, and rivulets feeding the stream had much to do with the survival of a turtle that was laying her eggs a half-mile away. So I decided to try to find another possible but as yet undiscovered species in the area.

Several years ago, I did find a rare and endangered blue-spotted salamander in the vernal pool in the unmanaged plot of woods on the northwestern side of my garden. But it failed to appear the following year, even though I used to go out on rainy spring nights to see if I could find it again. Unfortunately, by the time that this land sale was announced, salamander breeding season was over, and I would have to wait a year before searching again.

In the mid-nineteenth century, when New England agriculture was in serious decline, 85 percent of the land was open fields. In these abandoned tracts, trees and shrubs and other flowering plants, which had been dormant for hundreds of years in the soil bank below the cultivated lands, sprang up again. In most places, white pines quickly crowded out diverse species of hardwoods, but the Forbidden Forest retained the original (or at least very old) hardwoods. Thinking over this fact, I began to wonder if this small section of land could possibly be a glacial refugium, an anomalous ecosystem featuring plants that for some quirky climatic reason endured through the period of glaciation that 60,000 years ago covered most of the Northeast. Two or three such sites had been discovered in coastal areas not far from Scratch Flat, so I thought the notion might be worth following up. Unfortunately, I couldn't find any of the species that would be likely to occur in a refugium.

All this was a pipe dream, of course, but I survive on pipe dreams—my garden being the prime example. But there was one other very remote possibility.

The property lay within the boundaries of the original Nashobah Plantation. In a case in a nearby town, which also had one of John Eliot's sixteen Praying Indian villages within its boundaries, the town had ruled that any of the legally recognized descendants of these original inhabitants could make a case for ownership of a contested property. I thought I might try to contact some of the local Native people to see if they knew about this latest development project. But I discovered two problems. The northwestern boundary line of the original village of Nashobah was marked on early maps by a large glacial erratic boulder that lay at the very edge of the proposed development. Although the boundaries were a little vague, this meant that the development would not be on land that had been controlled by the Christian Indians. More to the point, the original inhabitants of the village had been members of the now extinct Massachusetts tribe, so there was no one around who could make a claim. Or so I thought.

Nonetheless, against all odds, I set out to lay a case and sent my possible challenges around to the local environmentalist community.

Not that I had to. Word was out, and the local land trusts and environmentalists were gathering for a fight. It turned out, however, that the lawyer for the developer was a man who was well known in the real estate community and who, reportedly, had never lost a case.

CHAPTER ELEVEN

THE FRENCH LACE ROSE

The rains that brought on the floods that spring fostered a fertile late spring and early summer. Things went well in the garden, but then they went too well—day after day of hot summer sun and no rain, and the drought set in.

But as I've said, there were worse climate-related events that season, which seemed to be building toward a record-breaking year on all fronts—historic wildfires, the worst drought in 1,200 years in the American Southwest, terrible floods in Bangladesh—not mention new studies indicating that the Arctic is warming four times faster than previously thought, which means that predictions about droughts and floods and rising sea levels will be coming sooner than expected.

What's more, things were not going well on the local front when the summer drought began. Weeks passed, and I continued my daily works-in-progress in the garden: tending new roses, keeping my frog ponds full, and constantly watering threatened trees and shrubs. I watched sadly as the garden continued to slowly dry out. The leaves of the mountain laurels and Mahonias curled and dried, the rhododendron leaves turned brown, some of the newly planted roses struggled desperately to stay alive, and

I began to dread next year's spring, when most of the damage of the summer drought would make itself apparent.

My troubles were minor compared to the heat waves in Europe. Flights out of London were delayed or canceled because the tarmac at Heathrow was melting in the record-high heat. Extreme heat waves continued throughout the summer, and later accounting determined that more than 60,000 people died as a result. In North America, there were water shortages and drying rivers, and fires in the West burned thousands of acres, creating critical air-pollution conditions, which authorities predicted would get worse if the droughts continued in the inevitable fire season in the coming year.

Nevertheless, the trees and shrubs in the well-watered Forbidden Forest held their own, and plants were flourishing in the carbon-rich atmosphere.

Then one morning I heard the dreaded backup warning beeps of a machine working nearby. I walked over to the outskirts of the forest and saw a group of men with hard hats and clipboards watching a backhoe dig a pit in the ground. I knew what that meant: a groundwater-level test to see if the water would percolate well—a sure prelude to development.

By this time the local environmental community was attempting to build a case to save the land, and had been joined by an unexpected ally: a land-use lawyer named Roger Moore. Earlier in his life, Moore had struggled in vain, as so many do, to save threatened lands. He grew so despondent about his losses that he decided to go to law school. As a land-use lawyer, he took to operating as a sort of mole, or fifth columnist, for the environmentalists: by hiring himself out to developers, he made sure that whatever housing project got through would preserve at least part of what the developers had intended to destroy. But he also worked, usually pro bono, for conservation groups. And now the good Mr. Moore had joined the case.

"No Trespassing" signs notwithstanding, I continued to include the woods in my usual circuit, still searching for rare plants or insects. As I hunted, I began to acquire a little more knowledge about the local insect population, hardly an area of my expertise. Not that a single species of ant could hold off a multimillion-dollar development project. But earlier that spring, during a so-called Biodiversity Weekend, there had been a huge ecological survey of the region in which experts and amateurs in a range of fields had undertaken a thorough accounting of local plant and animal species over the course of two summer days. E. O. Wilson, who was one of the prime movers of the event, had discovered a new species of ant on a hill not five miles west of Scratch Flat. This could have no influence on our cause, but it did prove, as E. O. himself had pointed in numerous publications, that we are destroying species of plants and animals even before we have been able to identify them. Who knew what undiscovered member of a little-known insect family might be living in the unique ecosystems of the woods and local fields in and around the Forbidden Forest?

I did manage to collect a list of well over a hundred insect species, including wasps, bees, hornets, flies, dragonflies, beetles, sucking insects, moths, and butterflies and seven familiar species of spiders, plus springtails, which I had seen there the winter before. But try as I might, I failed to find anything out of the ordinary.

By now the local planning board had begun reviewing the project, as had the conservation commission and other town agencies. A number of restrictions on the development were imposed, although none was out of the ordinary, and none would halt the destruction of the old farm and the woods.

The meetings dragged on. The locals rose up. Neighbors who had once shown little interest in nature or conservation issues or in local plants and animals joined the fight, spurred on by the infamous "Not in My Backyard" sentiment. Arguments thickened against the project. Lawyers battled. There were heated exchanges, and then, finally, the big day of decision came to pass: the day when the board of selectmen would set the final plans in place.

An ad hoc group had formed to encourage the town to buy the land as a part of an earlier town plan to preserve open space. But the developer was taking a hard line and would not agree to a reduced price. In fact, as the heroic Mr. Moore had discovered, he had upped the price far beyond what he had originally paid for the land. Nonetheless, the ad hoc committee members forged on and brought the proposal to buy the land before a special town meeting, which I attended as an observer.

It was a pleasant June evening, the kind of night when my wife and I would commonly enjoy a late dinner on the porch overlooking the gardens, listening to the meadow crickets and the last calls of the gray treefrogs and enjoying the moonrise over the scented air of the nearby flowerbeds. Instead, here I was in a brightly lit windowless meeting room crowded with attendees, with a line of grim-faced local officials shuffling papers on a long table in the front of the room and looking very businesslike indeed. And off to the side sat the developer with his bank of lawyers and his engineers and his toadies.

After a few closed discussions among the selectmen and planning-board members that no one else could hear very well, the moderator called the public meeting to order. The first part of the meeting was exceedingly boring. Via maps and charts and technical language, the planners reviewed the various restrictions. Then the developer, his lawyers, and his engineers explicated in tedious detail how they would meet said regulations and why this would clear the way for setbacks and curb cuts, lot sizes, access roads, drainage systems, elevations, slopes, discharges, and compensating wetlands.

This went on for nearly an hour. The room was hot and bright, and the assembled crowd remained silent, save for one older resident who, seemingly ignorant of meeting protocol, kept breaking in with arguments and pointed questions and repeatedly had to be silenced. The board members seemed to be familiar with him, as if he were a character who often showed up at meetings.

Finally, the moderator opened the meeting for comments from the public, not one of whom was in favor of the development. Roger Moore spoke first, and eloquently. His strategy was to oppose the whole development on a variety of technicalities; and when the lawyer representing the developer began sparring, the skilled Mr. Moore circumvented. He knew he would lose the case, but he kept the pressure on, relentlessly dodging and counter-charging. Then, understanding that a clear victory was impossible, he offered a compromise.

The local conservation agency, coupled with a town trails committee, had laid out a complex system of interlocking trails on town land, one of which was scheduled to run along the banks of Beaver Brook but had yet to be officially cleared. The new development would prevent completion of that plan. The committee had already been working with town boards to have part of the development limited so that the trail could be completed, but the developer and his lawyers and his engineers had balked. Now our heroic fifth columnist brought up the subject again.

Moore knew that, because of the existence of intermittent streams, bogs, vernal pools, and marshes linked to the brook, the developer was having trouble meeting the requirements imposed by the conservation commission on that section of his grandiose plan. So Moore asked, if the developer would not permit the trail to go across his property, would he not perhaps grant public access to a forty-acre tract of land on the southern end? That section abutted a twenty-seven-acre woodland owned by a local sportsmen's club, thereby meeting the desires of the conservationists to preserve at least a portion of wildlife habitat.

Moore's summary included an ethical appeal for the preservation of open space in our time. This speech was intended for the public attendees; he knew perfectly well that any appeal to save nature in the face of a worldwide decline in wildlife species would never touch the hearts, minds, or profit motives of the developer.

He began his summary broadly.

"We know that in our time we are sailing into a dangerous future of uncharted waters. The disputed realities of climate change are now upon us," he said. "The world's oceans are rising; 60,000–year-old glaciers and ice sheets are melting year by year; uncontrolled forest fires are fouling the air of West Coast cities, imprisoning children indoors for weeks on end, taking the lives of immune-compromised people and the elderly, threatening economically important species of plants and animals, and flooding the coastal cities of the world.

"And what can we do? There is only one thing we can do, and that is preserve our local forests and fields. Green plants, trees, shrubs, fields, marshes, and swamps absorb daily tens of thousands of tons of carbon dioxide. The complex diversity of local animals and plants that create the sustaining web of life lie before us right here on this land, offering breathing room, beauty, solace, and peace and quiet in a world of noise and the clamor of war and violence. What we save now, is all we will ever save. Which is to ask, if not here and now—when?"

He went on and on, adding more details about the values of open space, citing, as conservationists so often do, the outright betrayal and the bleak futures of our children and grandchildren and the fact that we are willing to them nothing more than a world in ruins. So powerful was his delivery, you might have thought he was charging the developer and his minions and the town boards with crimes against humanity.

After a closing recapitulation of his case for preservation, he paused and remained silent, as if in prayer.

"I rest my case," he said, softly.

His proposal was met with thunderous, sustained applause from the people. The developer and his lawyers and his engineers and his toadies remained stone-faced. The esteemed members of the boards shuffled their papers. The applause continued and then slowly died down. Last standing in applause was an older couple in the middle of the audience.

Then, one by one, over the next hour, the people themselves came to the podium in defense of the land, raising all manner of arguments for saving the whole tract. They spoke of the beauty of nature and offered all of the usual arguments about the importance of preserving wildlife in our time and the presence of birdlife (a point that I sorely wished I could address because I knew more about the species that nested there, including the rare blue-gray gnatcatcher, than did anyone else in the room). One man put up a very good argument for water conservation, arguing that a large development of this sort would certainly affect groundwater levels and inevitably leach excess nitrogen into the waters of Beaver Brook, thereby encouraging the growth of duckweed and algae that would choke out higher aquatic species. Another cited the role of trees in carbon sequestration.

None of these arguments held any weight with the developer. He and his lawyers and his engineers had heard it all before.

There is a pond in the middle of the tract that cannot by law be drained and developed, and the lawyer for the developer argued that this was an important contribution to wildlife preservation so there was hardly a need to do more.

He was countered by Moore, who pointed out that a pond surrounded by asphalt streets and driveways, not to mention manicured lawns maintained with pesticides and chemical fertilizers, essentially serves as an island, a limited biological community that will eventually lose most of its native plants and animals.

The meeting went on until it appeared that all of the opponents' cases had been heard.

"Any further comments?" the moderator asked.

No one came forward. There was only silence. Then a man who had been listening quietly at the back of the room raised his hand and slowly approached the podium.

He wore his smooth black hair in a ponytail, was dressed in a blue denim shirt, and had a necklace of bright stones draped over his chest. He identified himself as John Pashamet, a Native American, a descendant of the local Native tribe that had greeted the first Europeans.

He began his comments with a long diatribe recounting all of the sins that had been committed against his people since that fateful day in 1620 when the English "invaders" had landed. He described the destruction the newcomers had wrought among his people and to the land that had once sustained them. And, then having made his case, he offered a rational and imaginative argument for saving the tract of land.

"This land was once our land," he said. "Our people have lived here forever. And now what? It looks like the world you created will not last. In the space of four hundred years, it is breaking apart. And what you are proposing here today on this property is part of that destruction, yet another act of violence against the Earth that once sustained you.

"So I'm asking this: Why not create here, not yet another act of destruction of nature, not another cluster of houses, lawns, asphalt roads, poisoning the earth with chemical fertilizers and pesticides that kill our brothers and sisters, the animals—the birds, the bears, and all the wild animals of our nation, our former nation; another development that uses more energy and pollutes more air and water and now warms the very Earth and air. . . . Do not do that again, I ask. I mean, why don't you create here a symbol of hope, a symbol of what could have been, a standing forest and stream and meadow that will endure as both a memorial to what was lost and a message of hope for a new path for the future?"

He paused, nodding slowly, as if thinking of what more to say.

"Thank you," he said after a moment. "That's just what I think." Then he turned from the podium and slowly walked back to his seat.

After each of the previous speakers had finished their statements, there had always been a round of applause. After Pashamet's speech, there was only silence.

Then the older couple in the middle of the audience stood up and began clapping. A few more people rose, then more, and soon everyone in the audience was on their feet, applauding steadily and shouting, "Bravo," until the moderator began pounding his gavel, asking for silence, indicating, basically, that Pashamet's comments had been noted.

Not that his speech and our response did any good. The developer and his lawyers and his engineers and his toadies and also the selectmen and the planning-board members themselves sat grim-faced and unmoved.

Following that evening, there were subsequent private meetings involving the developer and the conservation and trail-committee members and the planning board. Journalists from local paper duly reported the progress, and in the end the development was approved. But the forty-acre block of wooded land at the southern end of the tract was saved. The Forbidden Forest would no longer be off-limits.

PART THREE

World without End

CHAPTER TWELVE

THE CARDINAL RICHELIEU ROSE

There was another official announcement of climate decline that summer. It appeared that, in spite of the Paris Accords and subsequent meetings of the UN climate committees, the use of fossil fuels had increased rather than declined. But at the same time, reports were offering some hope: statistics were indicating that alternative energy sources are now equal in price to or even cheaper than fossil fuels. The problem is that the powerful fossil-fuel industries are reluctant to change.

Meanwhile, a few solutions involving trees had been devised to counter the increase. Two or three companies had managed to genetically modify certain species to absorb more carbon dioxide than natural trees do, and these adapted species are being planted around the United States. Of course, as usual, there are drawbacks. In this case, the trouble has to do with biodiversity. The genetically altered trees are pines and aspens, and they are being planted to cover a great deal of land, essentially creating a farmlike monoculture. Environmentalists and ecologists worry that these massive plantations, as with the lands controlled by the timber industry, will create ecological deserts.

And yet they carry on, the engineers and the ecologists and the whole scientific community, surviving on the fixed belief that science and human ingenuity will save us. Which is true. At this point it's the only thing that can—or could.

My friend Toby Moffit, who is a graduate of the Massachusetts Institute of Technology and a biochemical engineer who works with cell division and other arcane research projects, told me not to worry. Something will come up, he said, and in the end humanity will survive and flourish.

I, of course, wanted to know exactly which *something* that might be, and he listed the usual technical advances in renewable resources such as wind and solar, also nuclear power and, at some point, fusion reactors, although that last goal, he admitted, in spite of recent advances, is a long way off. He told me that the owners of more than 100,000 solar- and wind-power projects were seeking permission to connect to electric grids, so large a number that grid operators were overwhelmed by the volume.

He went on to tell me about colleagues who are convinced that small nuclear-power plants can eliminate the need for fossil fuels and will not produce carbon dioxide. They are now developing models to build them. These are not the huge, complex installments that wreak havoc if there are accidents. If things go wrong, they can easily be contained, he said.

"What if none of these grand plans comes to fruition in time?" I asked.

"We'll get something new and, if all else fails, we may end up living underground," he said. (I'm not sure he was serious, but I listened.) "The very rich are already living in huge houses with furnished underground rooms and even swimming pools."

"And where will they get the water to fill these swimming pools, let alone to drink?" I asked.

"The sea," he said. "They will pump the seas inland to huge desalination plants."

"Where will they get the energy to run the giant desalination plants, much less export the water to the inland deserts?"

"All sources working together. There has never before been such a singular effort to get things done to halt climate change: amateurs, scientists, physicists, industrialists, financiers, kids, farmers, wastewater engineers. Plus, we don't know what new energy sources may turn up. Space-based solar power, for instance. We'll harness volcanoes and invent large-scale carbon-absorbing devices that will suck CO_2 from the atmosphere and turn it into stone. We'll block the sun with immense cloud-making machines. Better yet, drill down into the very core of the Earth for geothermal power. Projects of this sort are already being tested. Plus, there's hydrogen. In certain parts of the world, somewhere in Africa, I think, oil wells have brought up hydrogen, which burns with a clean flame and whose only emission is water, so we could search the Earth for hydrogen deposits instead of oil or gas. Retrofit existing power plants and even recycle the water vapor to supply clean water. The fact remains, there is no silver bullet that will save us; it's going to take a barrage of silver bullets, all working together. But you can hardly count the new systems nowadays, the new thinking that is going on, readjusting retirement portfolios to include green initiatives, creating living reefs of oyster beds to control future storm surges, and this is just the beginning. The race is on. There's hope."

"How long for these transformations?" I asked again.

"Yeah, I know, I know," he said tiredly.

Toby always was, *enfin*, a realist.

Another, only slightly off-center, friend of mine also believes in underground cities. He told me that in the past, in Cappadocia, in Turkey, there was a vast subterranean city that lasted for a thousand years or so.

"Toronto has a huge underground mall associated with two subways stations, plus there are other large underground malls around the world associated with subway systems," he said.

But he, too, was stumped by the fact that human life is dependent on natural resources. I told him that I had read recent reports indicating that, in some regions, vast areas of precious agricultural lands in certain tropical countries were under water, some feared permanently. I said that we still need soil, air, water, and light in order to eat.

He, too, surrendered.

There are other engineering solutions that can reduce carbon by using renewable natural resources, mainly trees. Climate activists are encouraging construction companies to return to wood rather than glass and steel and concrete. The production of steel and concrete are major contributors to greenhouse gases.

Researchers at the University of Maine have a program known as Green Energy and Materials that, with the help of huge 3D printers, transform wood waste such as brush into housing. The waste is ground into a powder, which is mixed with an organic binder to create a paste. Then, using a device rather like a huge toothpaste tube, it squeezes out the mix in layers to construct a building's foundation, walls, and roofs. The program plans to use the same material to create a series of windmills on rafts that would be anchored to the floor of the Gulf of Maine. They will be designed to withstand the high winds and erratic rogue waves that occasionally sweep through the gulf and will be located far off shore so that nobody on land will be able see them. There, they will also be out of the way of migrating birds.

But as with other innovations, the engineers and planners have not included time in their equations. According to most climatologists, the globe may be warming faster than human ingenuity can work.

Nevertheless, the race is on. I think it was my Dr. Pangloss Rosen, or someone like him, who told me the situation is similar to the U.S. production of ships, warplanes, and weaponry that began shortly after America entered the Second World War. It can be done. But we need the total cooperation of the U.S. government. Furthermore, any of these engineering solutions would have a better chance of succeeding if we could slow the creation of greenhouse gases. So far that is not working: levels are increasing year by year, but so are efforts to slow the process.

Engineering solutions notwithstanding, perhaps the greatest move toward reducing carbon dioxide and supporting action on climate change is the rise of citizen movements to influence change on both state and federal levels. There have been sizable demonstrations on both sides of the Atlantic, and many American cities and towns have established small citizen groups that are making a variety of efforts to battle climate change. These generally unnoticed local efforts come together across the nation on the unofficial holiday known as Earth Day.

I was actually involved in the movement to create the first Earth Day. In the late 1960s, I was working at a small environmental and nature center outside Hartford, Connecticut, when I learned that an environmental group would be coming to the city to organize some sort of celebration of the Earth. The group was looking for allies and venues and hoped to make it clear to the public that this is, in the end, the only planet we have and we have to take care of it.

I went to the organizational meeting, which had mostly drawn young people, some of them college students. The spokesman was a man from California wearing a taxi driver's cap and steel-rimmed glasses. He spelled out the idea of creating the holiday and outlined what groups could do to support the project, such as offer workshops and events about air and water pollution, pesticides, the dangers of nuclear power, and other topics. I, among others, agreed to organize and host such an event at the center where I worked and to contact friends, allies, and fellow travelers. It was a great success. We got far more participants than we expected, and we offered workshops on all manner of environmental issues as well as nature studies, trail hikes, an evening of environmental films, and even yoga and tai chi classes, which back then were considered a bit outré.

The workshop on nuclear power was run by a man I knew who worked at a local nuclear plant but was a sort of mole who actually didn't think nuclear power was a good idea, mainly because of the waste issue. I noticed among the crowd that day two men dressed in suit pants, white socks, and clean button-down shirts. They didn't seem interested at all in our various nature and environmental programs, only in the workshop on nuclear power.

I chatted them up later and asked in an off-hand way why they were so interested in that subject. It turned out they were from the company where the leader of the nuclear workshop was employed. As they phrased it, they just wanted to know what he had to say.

I heard he was later fired. But that didn't matter to him. He began working at another local environmental center, one focused on halting the development of nuclear plants.

The first Earth Day was a great success, and the holiday is still with us today. In Concord, Massachusetts, it is a major weekend celebration. The local arts and environmental center has used the holiday to emphasize our relationship with the natural world. Every year in winter, as a part of a craft program, children build large papier-mâché animal puppets. Then, on Earth Day, they hold a grand animal parade through the streets, led by Otter, their totem animal, and featuring replicas of salamanders, frogs, foxes, butterflies, and dragonflies. More recently, the center has split into two groups, one hosting lectures and art shows focused on climate change and other environmental issues, the other working at the grassroots level—educating children and holding outdoor events such as the animal parade.

I went to the event in 2022, as I had in years past, and discovered that, because of rain, the whole program had been moved inside the local high school. It seemed to me there were more local efforts than ever before, and they were all devoted to the climate issue. Even before I got into the main hall, I came across a man at a table promoting electric vehicles. It turned out that a state-supported program was offering financial incentives to anyone who bought one.

The main hall was a crowd of individual organizations, each offering one solution or another to deal with the issue of climate change. I went up and down the hall, checking them out. One was promoting a program to plant 250 trees in Concord to celebrate the 250th anniversary of the American Revolution. A group called Mothers Out Front was touting climate justice, and members were particularly concerned that the rising generation of children will be taking the brunt of future disasters. There was another tree-planting group and a natural lawn-care group seeking to reduce pesticide use. There were activities for children and a puppet show supporting Mother Earth. Two groups were encouraging bicycle transportation and expanded bike trails, though there were already several such trails in town, even one leading all the way to Cambridge. There was an interesting national student-run school program encouraging students to take action for the climate.

But the main focus that day was on two different groups that were fighting the expansion of a small local airport to accommodate more private jets. An earlier fight to halt the expansion had been successful; but like some modern version of the Hydra, an even larger plan had grown in its place. The environmental consequences of more private jets are legion. Along with carbon emission, airport noise is an increased concern. Exposure to chronic noise can increase risks of high-blood pressure, heart attacks, and strokes. Furthermore, once expansion starts and more jets begin to use the airport, the temptation is to expand even further to accommodate the new participants.

Interestingly enough, earlier that spring I'd met two pilots of private jets from the Midwest who were unaware of the controversy and cited the need for more landing sites. They happened to be part of a group that delivers critical medical supplies to sometimes remote sites around the world, so their support for such sites is understandable. But that sort of work was not the target of the anti-airport groups. They were arguing against private business jets and, worse, the planes of the super-rich who want to hop to, say, Martha's Vineyard or Cape Cod or Maine for the weekend.

At this time there was a similar protest taking place on the eastern end of the North Fork, on Long Island. High-end executives, anxious to avoid the notoriously heavy outbound summer traffic into the wealthy South Fork, were using corporate helicopters to carry them over Long Island Sound to their expensive villas, crossing over the quiet towns and vineyards of the more rural North Fork every Friday and Sunday. The locals rose up and this time were victorious. The helicopters now circle out around the easternmost point rather than cross over the towns of the North Fork.

I see a metaphor in this particular environmental issue that extends to land use and eventually to climate change. The super-rich of the developed world build their immense manor houses near urban areas and then fly off in their private jets to second or even third luxurious vacation homes in tropical venues, destroying the ecologically critical local mangrove swamps in the process of construction. Then, just to get away from it all, they build huge yachts the size of ocean liners. Although there is no documentation that I know of, the carbon footprint of a single wealthy family of this class must be huge.

Most of the visitors to Concord's Earth Day celebration that year were middle-aged environmental types, many of whom had been supporting conservation causes for years, some of them seasoned fighters who had spent time in the pens after being arrested at large demonstrations. But the people working the tables were younger, most of them local high school or college students.

I had witnessed this hopeful trend at family events in Washington, D.C., New York, and Virginia. One was a funeral, the others were weddings, and at all of these gatherings I had drifted into conversation with the younger crowd, some of whom lived on the West Coast. A few were in grad school at Stanford, one or two were just out of college and had jobs, and one was an authority on stratocumulus clouds (of all things), which are increasingly common because of climate change. One was working in London, attempting to get international companies to adopt so-called ESG protocols—that is, environmental, social, and governmental policies. Another encouraged power companies to install a specialized machine part that would improve fuel efficiency. A graduate student in economics was not hopeful about this effort. He pointed out that private corporations are interested in short-term profits, not the higher profits that occur over a ten-year period.

In connection with this, children of friends were already working in environmental and social affairs. One couple had helped to establish a group addressing the impact of climate change and racial discrimination in urban areas, issues that urban land-conservation efforts had traditionally not considered. They and their co-workers had set up workshops and lectures, were encouraging the replanting of empty lots, and had drafted people of color to join the cause. The group was also trying to instill a greater appreciation of nature among kids via a program that the Massachusetts Audubon Society was expanding in cities such as Boston, Lowell, and Lawrence. The goal was not only to encourage environmental awareness but also to share the deep history of the Native American cultures that had once lived on this land.

Their work reminded me of another group, the Boston Urban Gardeners. Founded back in the 1970s, it had worked to help establish community gardens in formerly overlooked and undeveloped vacant lots in low-income neighborhoods around the city. This group had had an interesting origin. During the 1940s and 1950s, there was a large influx of African Americans from the South into the northern cities. In Boston, many settled in the Roxbury neighborhood. One of those migrants was a man from North Carolina named Ed Cooper. Like many of the older Black people in the neighborhood, Mr. Cooper, as he was always called, had grown up hoeing turnips and pulling weeds on his father's farm; and when he left the South, he had vowed never again to touch a hoe. But contact with the earth, especially in childhood, becomes a part of the soul. The scent of mown grass or freshly turned soil summons memories and, after retirement, Mr. Cooper began to think of his past in a more positive way.

Roxbury was one of the more dangerous sections of the city. It had a high crime rate, some of the older people had been mugged, and there was a lot of drug use and gang violence. Near Mr. Cooper's apartment there was a thicket of high weeds strewn with litter, discarded needles, and dead cars. He got the idea that, with a little work, he could plant a garden there. Many other Black residents in the neighborhood must have had the same sense of belonging to the Earth, and they agreed to help clear the land and plant the lot with vegetables.

Mr. Cooper's garden flourished and after a few years came to be known as Cooper's Place. Some years later, a friend of mine, a garden journalist and photographer named Jerry Howard, discovered the garden and published a story about Mr. Cooper. The news spread, other community gardens were planted in empty lots, and, in the late 1970s, the disparate groups founded an organization called Boston Urban Gardeners, known by the fitting acronym BUG.

Eventually the Massachusetts Audubon Society took over the area where Mr. Cooper's garden was located and established, around his community garden, an inner-city sanctuary known as the Boston Nature Center, a sixty-seven-acre urban educational center for children and adults. Mr. Cooper was honored for his efforts with the Education Award, presented at the society's annual meeting.

Meanwhile, other towns and cities around the United States also began to establish community gardens. One, known as Esperanza Garden, sprang up in a New York City neighborhood with many Puerto Rican residents. Located in a dangerous area, the vacant lot was filled with detritus. So locals organized a workforce to clear the lot and plant a community garden. The site became a model for other gardens around the city, but over the long run it did not fare well. The lot on which Esperanza Garden sat was owned by a developer who was also a friend of the powerful mayor of the city, none other than Rudolph Giuliani. After allowing the garden to remain in place for seven years, he ordered the police to move in so that the developer could plow the garden under.

Not a good idea. The gardeners and the neighbors resisted.

News of the looming destruction spread throughout the Lower East Side. The state's attorney general, Eliot Spitzer, attempted to halt the project, and celebrities such as Bette Midler joined the demonstrators. The people erected a huge statue of a Puerto Rican frog, the coquí, which, legend holds, will resist attackers. The resistance fighters made plans to chain themselves to trees and cement blocks, and ultimately the New York Environmental Law and Justice Project joined the cause. But the mayor held his ground.

The protests mounted and the police were repeatedly held at bay as the attorney general's case progressed through the courts, but it all ended early one morning, when armed police moved in and began hauling away the cars that were blocking access to the garden. Then the bulldozers arrived. By 7 a.m. a crowd of protesters had chained themselves to items around the garden. But shortly after 10 a.m., officers began cutting the chains and carting protestors off to local precincts. Although the protesters had hoped to delay the attack until Spitzer's lawyers could argue their case in court, they failed. By midday the last of them was hauled off, and the gardens that had been providing food and air and consuming carbon dioxide for seven years were plowed under.

It should be said that efforts to improve city dwellers' access to nature are hardly new. In the mid-nineteenth century, landscape architects and park and garden designers began developing large country parks in the heart of crowded cities. The best known, Central Park in New York City, was designed by Frederick Law Olmsted and Calvert Vaux and opened in 1858. There followed a host of urban green spaces, including Olmsted's ambitious Emerald Necklace in Boston, Prospect Park in Brooklyn, and green spaces in cities throughout the Northeast and beyond. One of the inspirations for Olmsted and Vaux's work, and for my own garden, was the house and garden designer Andrew Jackson Downing, who published a gardening journal called *The Horticulturist* that offered new ideas and theories about landscape designs and rural subjects.

Since then, ideas about urban parks have evolved and spread throughout the world. Sometimes they have led to experimental and interesting projects, a movement that is collectively known as *ecological art*. Certain critics, most of them not art experts, are reluctant to use the term *art* for these projects, even though the people who design them are usually established and successful artists in one field or another. But no matter what they are called, the artworks themselves take an active role in improving and restoring urban environments, which now, among other positive effects, consume carbon dioxide.

One of these artworks, titled *Time Landscape*, was created by Alan Sonfist, who grew up in what he has called "the teeming jungles of the South Bronx," a district ruled by gangs and covered with concrete. He found a personal sanctuary in a deep ravine along the Bronx River (which has since been reclaimed and hosts a colony of beavers along with muskrats and species of formerly extirpated fish, birds, turtles, and frogs), and this childhood experience inspired his later work.

Time Landscape is situated on a 9,000-square-foot space that was once a trash-filled vacant lot on the corner of Houston and Bleecker streets. Sonfist cleaned up the garbage and then designed a mini landscape of trees, shrubs, and herbaceous plants such as gray birch, oaks, red cedars, native grasses, and other flowering species that would have been common on this plot in pre-colonial times. Using plants to construct a work of art was unheard of in 1978, when the project was created, but it has caught on, and other small plots have sprung up around the city.

Another ecological artist, Patricia Johanson, has created living artworks in both the United States and Brazil. In one of them, *Leonhardt Lagoon*, she transformed a stagnant polluted pond in a Dallas park into a thriving natural wetland, a safe habitat for native plants such as delta duck-potato as well as for native fish, turtles, and freshwater shrimp. Johanson included an organic walkway replicating vegetation that winds, vinelike, through the wetland. Now the public, especially children, come regularly to observe close-up the turtles and bird species, such as herons and egrets, that have returned to this wild oasis of restored environment in downtown Dallas.

The driving force of all of these artists is to create art that is actually restorative. For example, *Revival Field*, created by the artist Mel Chin and the agronomist Rufus Cheney, sits in the middle of a Superfund site in Saint Paul, Minnesota. It consists of corn, fescue grass, bladder campion, and pennycress laid out in a circular pattern bisected with paths that form the shape of a target. These flowering plants are growing in soil contaminated by industrial sludge that contains toxic heavy metals such as zinc and cadmium, hardly a place one would typically choose for a successful garden. But the garden was planted for what is known as green remediation. The plants that Chin and Cheney chose to feature are able to absorb without ill effects the heavy metals in the soils. After harvest they are burned in controlled conditions for biomass energy, and the heavy metals can be collected and sold. The soil left behind is purified.

We live with the great sword of Damocles hanging over our heads. Yet along with direct action and mediation, there are many ways—psychologically, at least—to get through this theoretically worst of all possible worlds. My personal choice is to take Voltaire's advice and plant a garden, but there are other, albeit less ecologically beneficial, means of getting by.

Late in the spring of that year I was out scything a section of the grassy mead that grows on the southern side of the driveway when a well-dressed couple came up the drive and approached me in a friendly manner. I knew exactly who they were—Christians of one sect or another, come to bring me the good news of salvation.

Unlike most of the people I know, I actually like to talk to Christian proselytizers and others who stop by here offering something that will help me live a better life. I'm not sure which sect these people represented, but they carried literature and began to talk to me about my faith and how Jesus will save me.

In all innocence, I asked, "Save me from what?" So we had a little chat about sin and forgiveness and how the Lord rules over the universe and all that. And then I began asking them about climate change and what's going to happen in the future, whether Jesus could save us from floods and droughts and all the other end-time events that scientists are warning us about.

Basically they told me yes and no. "The Lord's will be done," they said.

"Man cannot know the mind of God and his ways," the well-dressed man said. "He will save the righteous, those who turn to Him and accept Him into their hearts and minds. Those he will save."

"What if you don't accept him into your heart and mind?" I asked.

Almost apologetically, the man, who had introduced himself as George Mason, answered all of my questions. His assistant, a pretty woman with a tired aspect, as if she had endured a rough life, remained silent, nodding gravely at everything he said. George explained gently that if I had no faith—no Christian faith, that is—I would be left behind.

"This climate-change thing, you see, is part of God's plan. At the end of time, he will sweep up the righteous and save them, but all the others will be left behind."

"So what?" I asked. "I like it here in the garden."

He smiled condescendingly and hesitated again. He didn't want to be cruel, but he explained gently that, along with everyone else and, I presume, all the animals, we will die.

"Well, we're all going to die," I agreed. "Even old turtles die."

"I mean forever. The saved will return to live again in a better world."

Then I asked them about extinction: "How about all the animals who are going to die during the droughts and the floods and fires that God has sent down? Can animals get into Heaven?"

The two of them contemplated this before answering. "Animals don't have souls like humans do," George said. "So they can't get into Heaven."

"Where do they go?

On this he was clear. "They just die," he said. "They go back to the Earth."

Then I sprang another unanswerable question: "If animals can't be saved, why do they have to suffer sickness and pain? Why doesn't God just let them have a comfortable life and then let them just keel over in a painless death?"

"We cannot question the ways of the Lord," George said again, apologetically.

Then I asked if they believed that climate change is going to alter everything in a matter of decades.

"Who knows?" George said. "But it is perhaps in God's plan to destroy the world so he can remake it. As I say, a certain number of believers will survive, and they will repopulate God's New World. He has done this before, as you may know."

"The flood?"

"Yes. Noah's flood," he said.

Hereupon they offered me some literature describing the hope of salvation if I would but turn to their religion. I accepted their offer and took the pamphlets.

"Anything to get out of the mess we've created," I said.

There are times when I often envy those of good faith. Hell and high water don't seem to affect them. I asked if they had found peace of mind in their faith, and they both enthusiastically said they had. Even in troubled times, they can turn to Jesus.

At this point, the woman finally spoke up. "You don't know my story," she said. "I was lost, but now am found. I had a rough life, drugs, sex, too much drinking. Outwardly, when drunk or high, I was happy. But I would wake up broken. All the world was black. I grew up in Florida, and one morning after a wild night which I couldn't even remember, I woke up and found that I was lying by a slow-running stream not far from my home. I guess I had passed out. Anyway, I woke just before dawn. People near us had chickens, and I think their rooster woke me. I sat up feeling, as I so often did in those days, 'What's the point?,' and then I heard the rooster call again. I'm like still down in the dumps, and then it called out again and something broke. I don't know what it was. I started crying and couldn't stop, just sitting there in the swamp, bawling. Then I realized why.

"See, I came from a religious family, off to church every Sunday, Bible school and all that. But then I strayed. Boredom, maybe, teenager, and I'm going out with these kids who like to have a good time, drink cheap whiskey, smoke dope, and then other drugs, and I'm like having sex with guys I don't even know. But I start waking up unhappy. Only get happy with my friends.

"So there I am, sitting in the swamp in the dawn light, and I hear a voice call my name. I turn around and there's no one there, But the voice spoke. I heard it clearly, like the person was right behind me, a man's voice, a kind voice, soft. He says, 'Before the cock crows thrice ye shall betray me.'

"That did it. I thought at first I was tripping, but then remembered that line in church and I'm thinking—that can only be the voice of God talking to me."

George interrupted, "She's talking about the scene in the garden of Gethsemane when Jesus tells Peter that he will betray Him before the cock crows three times."

"Yeah, right, Jesus talking. So then I started crying again and I can tell you, those were tears of joy. I never had no experience like that one, and I been on a lot of weird trips. But this was real. I had betrayed my Savior and he's offering me salvation. Like they say, I felt like I'm born again."

It was quite a story, but I had heard it before. As I mentioned, I have the curious habit of actually engaging with those religious proselytizers who stand around in public places preaching the Word. One of the common stories I hear is from drunks who've found God.

That's one way out, I suppose.

There is another, more pragmatic group of Jewish and Christian believers who also see God as a prime mover, but their God tells them to take action on climate issues. The activist Bill McKibben is a churchgoer who has worked closely with Jim Antal, a Church of Christ minister, whose sole mission, starting in the late 1970s, has been to encourage climate action. The two have attended hundreds of demonstrations, sweated together in crowded police wagons, and spent time together in jail because of their climate action.

It appears that there is there is a rising tide of support for the environment among people of faith. I learned this when I was editing the environmental journal *Sanctuary*, which focused on a single theme in each issue. One was titled "Ecology and Theology" and dealt with the role of religions in environmental concerns. It was the most popular issue we ever published; churches and synagogues from all over the Northeast ordered copies, so many that we ran out and had to reprint. The liberal Jewish magazine *Tikkun*, which actively works for environmental sanity, among other causes, wrote to us in support, and the Catholic diocese of Worcester, Massachusetts, ordered so many copies that we had to limit the number we could give away.

Church historians have pointed out that religions have often been at the forefront of social change. As far back as the sixth century, when the farms and forests of Europe had been ravaged by the excesses of the Roman Empire, the newly formed Benedictine monks, called by their God, began replanting local forests and working to improve the watersheds of streams and ponds. In the late eighteenth century, the Quakers and other Protestant sects in England and the United States began the abolition movement that eventually ended slavery. One hundred years later, the anti-Nazi German pastor Dietrich Bonhoeffer introduced a new version of Christian ethics that involved dissent and resistance. Of course, his new church was shut down; he was arrested by the Gestapo and executed in prison just before liberation.

Not all of the faithful are driven by God. Both Ralph Waldo Emerson and Henry Thoreau had controversial ideas about the nature of the deity. By contrast, that other great conservationist, John Muir, was deeply religious. (In spite of his faith, Muir was also said to be a racist.)

As far as I know, Rachel Carson was not particularly religious, nor was Marjory Stoneman Douglass, the driving force behind the establishment of Everglades National Park.

Thinking this over, I remembered conversations with an old friend of mine, a politically active Unitarian minister named Bill Henderson. I'd met Bill through his son David, who told me one day that, when he was about ten, his father had announced that they must take a walk together.

"He took me up to a path on a high hill above a series of fields with grazing cattle. He was quiet most of the way, but when we got to the top overlooking the green pastures, he halted and surveyed the landscape.

"'Son,' he said gravely, 'I have to tell you something.'

"His tone was such that I thought he was going to tell me he and my mother were getting a divorce, or that he was ill with some incurable disease.

"Instead, he remained silent for a few seconds and stared out over the fields. Then he turned and took my arm, facing me.

"'Son,' he said, 'THERE IS NO GOD.'"

This might explain Reverend Bill's inspired worldly sermons on political activism and his constant community service, plus his periodic arrests for civil disobedience. He was part of a local group of older activists in the town, many of them his parishioners, who formed the equivalent of a local leftist cell that protested any destructive new developments, such as the ruination of Charlie's fields. Most had been involved in the civil rights protests of the late 1950s. Some had been at the bridge on Bloody Sunday in Selma, Alabama, during Martin Luther King Jr.'s famous march to Montgomery.

Earlier that spring, I had fallen into a discussion with one of my friends, a professed Buddhist. I asked her what the Buddha would do in the face of climate change. How are we to save ourselves from all this?

"I'm not sure," she said, "but I've read that the Buddha says that if we take right action for the benefit of all, we may not be able to prevent disasters like climate change but we can learn how to live comfortably with the realities—the extinctions, the political upheavals, the economic disruptions, and all that."

"How's he going to do that?" I asked.

"He's not going to do it. The Buddha doesn't do anything. *You* have to do it. But I think he would encourage taking right action."

"Which means?"

"As I understand it, it's a little like the Ten Commandments, a series of paths, one of which is to eliminate cravings for material goods or fame or wealth, which in the end are never satisfying anyway.

"I think it also involves making a living in a system that does no harm. The Buddha would say you have to review the course of your life. Look within. You've got to change your consciousness from self-centered desires, material desires, quests for power or recognition or fame or identity and turn toward a life of altruism and compassion."

"Let me ask you this," I said. "Do you think those currently in power, the world leaders and the unhappy few who own immense yachts and five houses in favored climes, the ones who hold 99 percent of the world's capital—do you think they would ever surrender their world and wealth, look within, and give away all their money and follow the path of righteousness all the days of their lives?"

She laughed and shook her head. "Not likely," she said. "But there are certain traditions in Hindu cultures in which, after a certain age, having acquired wealth, men (not women, as far as I know, not mothers) will give away all their money, leave their families, take up the beggar's bowl, and wander off seeking enlightenment. In fact, the Buddha himself did that."

"Sounds good," I said. "That may save them perhaps from despair or climate anxiety, but it won't help slow the advance of global warming."

"Well, according to the Buddha, life is impermanent anyway, and living is just a process of flow. He would say there are a lot of ways to cope with radical changes such as global warming, but learning to live with some degree of inner comfort involves living in the present, what some people call the 'now moment.' This can free us, they say, from fears of endings, including death, and also radical environmental change."

She reported this with the air of a student, not a confirmed believer, and in fact she'd been raised in a Jewish household. But she clearly wanted to believe in some spiritual path other than the one in which she'd been raised.

"Also, I heard an interview with the Dalai Lama recently," she added. "He did have an answer. He said we should all become vegetarians. Cows emit a huge amount of methane, a greenhouse gas, and require huge tracts of monocultures to grow their feed."

I told her I liked the sound of that. But how many Wall Street brokers actually worry about radical environmental changes? And as far as I know, many of the rich and powerful magnates are fond of steak. And even if they gave up meat, how long would it take to eliminate milk and meat dependency from the western diet?

Nonetheless, I had to agree that it was a good proposal, not unlike planting a garden. Both make a statement about the way the world should be.

I decided to check up on these religious solutions and attitudes to the climate issue with my atheist friend Dr. Pangloss Rosen, who had also been raised Jewish but now takes comfort in the study of history and what he always terms "the benign indifference of the universe," (not an original idea, I don't think; he was quoting some philosopher).

As far as the history of religions is concerned, he said, there must have been some strange vibration in the air during the fifth century BCE. Buddhism arrived on the world stage in that period, which, interestingly enough, is roughly the same period in which rational Greek philosophy, Jainism, and monotheistic Judaism began. It is also roughly the same period of the written version of the great Hindu text, the Bhagavad Gita.

Along with a number of other scholars, the historians Karen Armstrong and Lynn White note that the rise of Judeo-Christian monotheism established a clear break with earlier Hellenistic and Egyptian religions because they set humankind at the center of creation, apart from and above nature. That break, according to White's famous 1967 essay "The Historical Roots of Our Current Ecological Crisis," published in *Science*, was caused by the advent of monotheism and an indifference to the profane natural world.

According to White, the differing attitudes toward nature in eastern and western cultures are revealed in their traditional artworks. In the vast mountainous landscapes depicted on ninth-century Taoist scrolls, nature is dominant. Set against wild mountains and dramatic cloudscapes, a tiny hut with a sole human figure appears on a lower corner. In contrast, the paintings of the Renaissance feature large portraits of men and women and Christian religious scenes, also interiors, with a tiny representation of nature (usually cultivated) outside a window or in the far distance. That, in White's view, represents the core of the environmental issue.

White's essay was heavily criticized and not only by Christians. Moreover, his own views evolved. In a later essay, he admitted that the Buddhist culture had also quite effectively destroyed wild nature. Contemporary Chinese culture, he wrote, is not much better. I read later that Mao Tse-Tung had once introduced a campaign to kill all of the nation's birds in order to protect crops.

I described my various religious encounters to Dr. Pangloss, and, as was his wont, he had a lot to say. His disbelief in the Almighty did not hold him back from an interest in the history of world religions.

Like other scholars of history and religion, including Lynn White, he agreed that the Judeo-Christian monotheistic, anthropocentric, God-proclaimed subjugation of nature marked the end of earlier religions and cultural traditions that had accepted nature as an integral part of human existence. But he placed the break with nature much farther back in time; and unlike most scholars of ecology and religion, he therein found reasons for hope.

He agreed, along with most of the other critics, including Karen Armstrong, that the Paleolithic hunters had presumed a kinship with the wild earth, which was in their view also sacred. They nurtured that kinship spiritually and celebrated the wonder of their existence, along with all of nature. Although Dr. Pangloss acknowledged that this view of the web of life had since been torn asunder, he argued (citing many sources, as usual) that a shred of the ancient bond carried on, even after the monotheistic split, and in fact still exists.

"You look at a group of children at play without adult supervision," he said. "They will form bands and build forts and hideaways and places apart and defend themselves against imaginary bands of alien tribes. You will notice that girls and boys alike will instinctively take to climbing trees, thereby reaching back to the very beginnings of the earliest hominids. The natural world is their home. They will not actually study nature unless they attend some adult institution—a camp or a school course. But the very study of nature indicates a separation from nature, does it not? Children accept the natural world as it is without judgment. They even have a primordial fear of predators—lions and tigers and bears—which have not posed a threat to most human societies for thousands of years."

I had to argue with him on this point because here, in the industrialized cyberworld, free play outdoors among contemporary children is almost extinct. Most of their contact with plants and animals is experienced through films and TV shows about wildlife in far-off places such as Africa. In addition to its issue devoted to ecology and theology, the journal *Sanctuary* had published another best-selling issue—this one about the loss of free play among today's children. Structured outdoor programs, mobile devices, and outright fear of nature in the form of ticks and mosquito-borne diseases have subsumed everyday contact with the tangible world.

But he persisted: "The connection is still there, though, even in adults. Just regard some well-dressed businessman on the streets of Boston, striding along and attempting in exasperation to open one of the infernal packaging impediments of our time in order to get at the peanuts within. What will he do? He will tear the flesh from his recently killed prey: he will tear the bag open with his teeth."

"People still mate," he said. "Human females bear children and nurse them. Their children are born wild; they must be tamed and trained to live within the traditions of the hunting band they were born into. The males—and nowadays, true to the roots of our Paleolithic animal natures, also the females— are the hunters. They work in groups and make plans to hunt the bison and the woolly mammoths of our time."

His reference to women's acceptance into the contemporary workforce sent him off onto a long tributary discourse about the role of the female in Paleolithic hunting bands. "Archeological evidence indicates that both sexes were hunters and warriors," he said.

That, in turn, sent him off to the Scythian horse queens, the warrior women who terrorized even the warlike Greeks. "The Amazons were not mythological beings," he said. "They were Scythian women warriors."

"So where's the hope in all that?" I asked. "The modern hunters, men or women, have managed to destroy their hunting grounds and have driven their prey to extinction. I sometimes think the only guilt-free human beings left on Earth are the diminishing tribes of true hunter-gatherers. They may have some weird rituals and the like, but at least they're not harming the Earth with their way of life."

"Yes," he said. "Yes, but there's still hope. The ancient bond with nature was and is still with us. You can read about it between the lines, even in the stories of world religions and literatures."

"Their religions are what caused the break," I countered. "Their literature backs it up. Look at the *Epic of Gilgamesh*: the heroes kill the forest guardian Humbaba and cut down the cedars of Lebanon that he protected."

"But it is there," Dr. Pangloss said, "even in the foundational stories of the Greek rationalists and the roots of western cultures. Who saved Psyche from death when her mother, Aphrodite, attempted to kill her by posing impossible tasks?"

"Don't know," I said.

"Ants!" he nearly shouted. "Mere insects!"

He told me that early Taoist poets and scroll painters in ninth-century China had managed to accomplish this transformation. Many of them had fled from their highly structured bureaucratic society and had retreated to the frontier of the wild karst mountains of Guilin. He cited the poems of Li Po, Tu Fu, and Wang Wei, with their deep attention and connection to natural imagery, and the artists of the period, with their images of snow, of rising flocks of egrets, of great waters and mountains, of crickets and frogs, lizards and blossoms.

This deep union with nature carries on, he claimed.

I told him I had read about Lynn White's revised essay, which argued that the Taoists and Buddhists had also managed to destroy the Earth.

Dr. Pangloss didn't buy it. "You do know, do you not, about the significant interplay of man in nature in the story of the great Chinese classical philosopher Zhuangzi, who dreamed he was a butterfly? When he awoke he could not be sure he was Zhuangzi dreaming he was a butterfly or a butterfly dreaming he was Zhuangzi.

"Skip forward a few thousand years, and there it is again in William Blake, who wrote that he couldn't tell the difference between a fly and himself. Or the nature writings of Henry

Thoreau, or Emily Dickinson, or the violent radical poetry of
Robinson Jeffers, who seems to be coming back into favor in our
time. My point is, the seed is there. It's like those wheat seeds
archeologists found in some tomb from ancient Egypt. Even
after 6,000 years of dormancy, archeologists sprouted them."

He ran out of breath at this point and looked out at the slow-
moving waters of the river system that had inspired so many of
the Concord writers. Uncharacteristically, he remained silent
for a minute or so.

"All the waters of the world come together," he said finally,
apropos of nothing.

I'm not sure how this comment connected to his earlier arguments, but his main point that afternoon was that the real solution to climate change and the extinctions of the Anthropocene
is to return to our original wild Paleolithic nature—or at least
to its underlying philosophy of oneness with nature.

It was rather odd to hear this philosophy, coming as it did
from the likes of Robert Rosen, an inhabitant of windowless
library stacks, a man who spent most of his days indoors, reading and writing and lecturing on a variety of subjects. A man
who wears thick glasses, is slightly overweight, always wears a
necktie (although always slightly skewed), is fond of whiskey
and not opposed to eating red meat. Meat, he used to say, was
the primordial food of our wild ancestors.

In another conversation we had a few weeks later on the same
subject, he did admit, along with some engineers and ecologists
and most climatologists, that it is perhaps too late and there may
be nothing we can do at this point.

Another good way to get through the psychological problems
created by global warming is to deny that climate change exists,
a position that was common until recently and is still held to be
a hoax by the uninformed and certain powerful political figures.

Currently, there are many in this country and probably the world around who accept the reality of global warming but just don't think about it. Only about 8 percent of the American population is actively working to do something about climate change

Then there are also those who are well informed about the science but take the long view. The late evolutionary biologist Lynn Margulis, whose theories revolutionized thinking on the origins of life on the planet, was one of these. She was also a great supporter of James Lovelock's Gaia hypothesis, which holds that the Earth can be viewed as a living thing, a single self-regulating organism. I interviewed her once while I was researching climate change for one of the journal issues I was preparing. She didn't seem too worried. "Climate change is bad for real estate," she said. "But the Gaia will endure."

I had a scholarly acquaintance, Skip Lazell, who was an island biogeographer and knew a thing or two about rising sea levels and mass extinctions and didn't seem to worry either. Lazell and other island biogeographers were researching the diversity of plant and animal life found on islands and how distance from the mainland or the size of the island affects their biodiversity. Generally, before the onset of the last glacier 60,000 years ago, the smaller islands were once mountain tops. Then, as the seas rose, the peaks became islands and over time were populated by species that were different from their mainland relatives.

Years ago, Skip was searching for rare species of reptiles and amphibians on islands in the South China Sea. I joined him there on an assignment while he and his crew scoured the islands just off Hong Kong for evidence of new or endangered species. They were attempting to do a biological survey of two of the islands that were about to be destroyed to make way for the new Hong Kong airport.

This was not the most comfortable overseas assignment I had ever been on. For one thing, the group felt it was necessary to go out at night into a large marsh on the island of Chep Lap

Kok, where, among other species, the krait, a highly venomous snake that hunts at night, was common. One bite and you die in ten steps, as the folklore says.

In the evenings, after days in the field, I would grill Dr. Lazell on his work, asking specifically about the problems created by climate change, given that many of his study sites would soon be under water.

But Skip tended to think in millions of years rather than in decades or even centuries, so he viewed things in perspective. "This has happened before," he said. "It'll all come back."

I couldn't help but think of George the Christian, the man I had talked to in my driveway, who'd said more or less the same thing.

Each to his or her own religion, I thought.

Concern about climate change has led to a new field of scientific study, popular trade books, academic treatises, theoretical engineered solutions, and other strategies that offer at least a glimmer of hope for a slowdown in the inevitable progress of warming. I have three friends, all originally trained as structural engineers, who years ago switched fields and began designing solar-powered houses or founding solar companies. A number of environmental writers, including Bill McKibben, who early on began writing about the conservation of nature, have since switched to crying out specifically against climate change. A certain percentage of people in western nations have reacted to climate change by attempting to live low on the food chain, recycling and composting, cutting out red meat or becoming vegetarians, driving electric cars, and eschewing air travel—doing all they can in their personal lives to at least absolve themselves from blame.

Among them are a number of nouveau Thoreauvians, some of them my friends and associates, who have taken to the woods, cut themselves off from the grid, built their own cabins, tried to raise their own food, and attempted to avoid living lives of "quiet desperation"—as, according to Thoreau, most people do. Not all these off-the-grid refugees took to the woods because of climate change. Most left the modern world in order to live deliberately, following the model of Thoreau. But by doing so they dramatically reduced their personal contributions to carbon loading, although this is a very small and mostly existential statement. The billion-plus people in Africa and India are not responsible for carbon input either, but they suffer the most when droughts and floods occur.

Compared to these new escapist pioneers, my chosen method of dealing with the climate issue is perverse. Roses require a lot of nurturing and careful growing conditions. Why should I go about cutting down native trees, digging in the natural soils, amending it, and seeding it with plants originally from Mexico, India, the Himalayas (rhododendrons), the South Pacific, and (accidentally) alien plants from Europe and Asia? Building a shelter from native trees (all of the wood for the so-called Vicarage came from local trees, milled at a sawmill less than a mile from my property) and trying to live low on the food chain is one thing. But including non-native plants in my garden is another. So is cutting grass and pruning shrubs and flowering trees, for that matter.

We all survive by altering the natural world for farms and pastureland. But relatively recent changes such as the so-called Green Revolution and the rise of huge industrialized farms and feedlots and massive chicken and hog farms that require herbicides, pesticides, and artificial fertilizers, have destroyed native species of plants and animals. And the operating machinery is a huge producer of greenhouse gases.

Small exurban organic farms, such as those of Scratch Flat and elsewhere outside of Boston and other big cities, are far more benign.

❀

Climate change is already affecting segments of society in the United States and around the world. Yet there are a lot of seemingly rational reasons for people to conclude that it is not as dangerous as the established scientific community claims. This position is perhaps understandable.

For the past thirty years or so, lobbyists for the fossil-fuel industry, media moguls, and some politicians have been questioning the science, attempting to convince the public that the world's various, unique, and record-setting disasters are not related to climate. Every winter, when there is a major blizzard somewhere or a prolonged cold snap, the flags of doubt are raised, and people go around saying, "So much for global warming."

There is, of course, a difference between weather and climate. A blizzard or a cold winter here and there is weather. Melting glaciers, early plant blooms, and a slow but determined rise in sea levels and worldwide temperatures are climate change.

Indifference to the situation is one thing, but outright active denial of its existence is quite another, especially when it comes to those who hold political and financial power. One study of the world's five largest publicly owned oil and gas companies indicates that they've spent about $100 million in attempts to block climate policy studies or at least slow them down.

But that may be changing. Starting in the early twenty-first century, groups of socially conscious investors began ranking large-cap corporations according to their social and environmental standards. The concept took hold and grew and has now developed into a new and divisive political issue, generalized as the ESG standards I mentioned earlier. The political right wing dislikes ESG; liberals support it. Cynics, such as Dr. Pangloss Rosen, question it.

"The idea is good," he told me. "If, for example, some economic *deus ex machina* were to descend and the politicians and the corporations and their investors were to accept and carry out this ESG idea, we might have been able to create actual environmental changes that would have slowed—but not stopped—the advance of warming. The idea has been floating around as an econometric model since the 1990s, and a few semi-responsible corporations, driven by socially active investors and their financial advisors, have accepted the idea. I personally have a financial advisor who deals solely in so-called green companies, and I've held my own, as you can see."

His current retirement home is not inexpensive.

He went on to say (with caveats) that corporations are beholden to their investors and that currently responsible investors are checking the environmental position of their companies. Even some fossil-fuel industries have been responding. A few years ago, activist investors in ExxonMobil launched a campaign to replace four executive board members who were reluctant to accept the terms of ESG and were thereby putting the long-term success of the giant company in jeopardy. Moreover, the leader of the asset-management firm Black Rock announced in a letter to executives that climate change would soon fundamentally reshape financial systems around the world. This statement, coming as it did, from the richest of all the asset-management firms, turned out to be a watershed moment. Corporations began to react.

"But," my Dr. Pangloss continued, "you have heard, have you not, of 'green washing'? Energy companies and even some nations that have agreed to reduce carbon releases outlined by the Paris Accords claim they are cooperating when, in fact, they are not. This is especially apparent in ads put out by fossil-fuel companies boasting of all their green advances, which are not untrue but turn out to be only a tiny percentage of their overall expenditures."

He went on to assert that what the fossil-fuel industries say they are doing is one thing. What they actually do is another. Furthermore, he said that, although (in 2022) the federal administration may have been attempting to cut back on fossil-fuel support and encouraging green technologies, the United States is still exporting oil as if we were not all living on the same threatened planet. He told me about a piece he had read recently claiming that a large nonprofit organization that specialized in planting carbon-sequestration forest lands was not exactly complying and furthermore was uprooting local Indigenous populations in order to plant the new trees.

I broke in at this point to declare my support for the growing green movement. Locally, the Concord Art Center and the Umbrella Arts and Environment Center had recently mounted shows dealing specifically with climate change. The artworks had centered on the positive changes taking place at the community level around the world, from the Arctic to the tropics. Plus, internationally, a large number of nongovernmental agencies are supporting gardening solutions in Kenya and other African nations and are working to help people in flood-threatened regions in India and Pakistan. I had also read a recent story about Indian villagers who were working independently to redesign the fields and houses in their villages to make them safe in the floods that will inevitably come. The article pointed out that these developing societies do not need news stories, books, or articles to help them grasp the destructive aspect of a new climate. They live in it.

I agreed with the good doctor that disbelief and indifference endure and that the right wing (in the United States, at least) is opposed to the "woke" and to the ESG drift. But I told him that the real hope for change is now coming from people under the age of thirty, some of whom are too young to vote but not too young to bring lawsuits against their indifferent elders who hold the seats of power.

Dr. Pangloss listened, but I could see he was gathering a counter-argument.

"It's a dark and a dreadful future they face," he began. "I agree they are trying. But look at what they're up against: a vast worldwide shift to the right. And in spite of the scientific facts and the reality of a disatrous future, those in power and the financial world at large will continue to be driven by the black horses of the dark god, Mammon. That is the systemic feature of capitalism. The powerbrokers are determined to stay the course, encouraged by the 1 percent and their international financiers and their politicians and corporate heads, including a past president of the World Bank and many of them still claiming that climate change is a hoax. It's business as usual and damn the torpedoes."

As usual, he used a historical analogy to make his case: he pointed out that the institution of slavery in the United States was the base of the nation's thriving economy in the early nineteenth century. The South argued, not without evidence, that "cotton was king" and that, if the abolitionists took away the region's enslaved laborers, the American economy would crash.

"You can see, then, why the fossil-fuel industries must carry on. World economies are enslaved to oil."

He then offered up a litany of the arguments for inaction:

- Climate is cyclical, a natural process, which is true (except it was not human-caused, as it is now).
- The climate models are unreliable and too sensitive.
- The carbon dioxide measurements are wrong, or the levels are so small they cannot be blamed for worldwide temperature increases.
- Halting climate change is too expensive. (Never mind, he added, that the U.S. fossil-fuel industries are getting well over a billion dollars in annual subsidies and tax breaks. And never mind that the costs of damage from climate change run in the billions of dollars.)

+ Cold winters kill more people than heat waves do
  (though this claim does not take into consideration
  the destruction of property and the resulting refugees).

+ The United States cannot take action because
  other countries are not taking action (though the
  United States, along with China, bears the greatest
  responsibility for carbon emissions).

And finally—and this after one hundred years of global warming—

+ We should not rush into changing things.

After another very long diversion into the Paleolithic cultures,
my Dr. Pangloss offered a summary: "Unfortunately—or
fortunately, I suppose—we, the late-coming Cro-Magnons,
can adapt readily to any adversity and any environment. Face it:
900,000 years ago, in the middle of the Pleistocene, the hominid
population plummeted to 1,280 individuals, from a high of about
100,000, and then recovered. And therein lies the problem.
Deserts burn, the forests burn, the rivers dry up. The coastal cities
sink beneath the great seas; death-dealing heatwaves sweep the
globe. Fires, crop failures, massive extinctions of economically
critical species, plagues, poisonous air: we've seen it all before,
and what do we do? We adapt and survive. It's easier to just give
up and think about it tomorrow.

"And so we beat on. . . ."

CHAPTER THIRTEEN

THE LAST ROSE OF SUMMER

Climate change grinds on slowly, but it grinds exceedingly small. The shifts are barely noticeable over the course of a single year, save for the year 2022, which experienced a newsworthy climate-related disaster at least twice a month. Yet these shifts were always apparent to gardeners and other weather watchers who had followed such things across decades. Notably, among other data, climatologists tracked warming trends using the detailed records of English gardeners and the weather-related, often meticulous, diaries of the wives of American and British farmers.

In my part of the world, the detailed natural history journals kept by Henry David Thoreau are another indicator.

In his time, Walden Pond used to freeze over in late November or earlier. Now, in some years, the pond stays open all winter. Thoreau's detailed entries document the flowering dates of plants over the course of his short adult life, including his accounts of first and last frosts. These records have become valuable to local climate researchers, in particular to Richard Primack of Boston University, who with his team tracks the advancing flowering times of plants by comparing them with Thoreau's dates.

Even more interesting are Thoreau's detailed records of specific environmental activity, such as the fall of light on Walden Pond on certain days and the silver gleam of light on the underside of certain plant leaves. As far as I know, the only other person who paid attention to this sort of thing was that other quester for sanctuary, Vincent van Gogh. His accounts of atmospheric conditions and the varying colors created by the fall of light on plants appear in the letters that he regularly wrote to his brother.

When I first moved to the place that became Sunny Bank Garden, we used to get a light frost around September 18 and a killing frost around October 5. By Halloween, the nights were chill, and in some years children had to wear winter coats over their costumes. Nowadays, we generally get a light frost around Halloween and a killing frost toward the end of November. One year not long ago, hard frosts did not arrive until mid-December and the rue and other hardy broadleaved plants were still green in January.

Not to rant yet again against the current decline of the environment, but among the other changes I have witnessed locally over the years are some major contributors to climate change—mainly the construction of new highways and housing developments in this formerly agricultural town.

Just before I moved to Scratch Flat, I was living on Martha's Vineyard, which (back then but no longer) was so quiet in winter that the drivers of passing cars would wave to one another. Life on the mainland seemed to be moving at a higher rate of speed than I was used to, and I had trouble adjusting, even though the five surrounding farms of Scratch Flat offered a semblance of rural life. The only traffic jams back then were caused by the slow-moving hay trucks that blocked the Great Road in late June. I was told that, not long before I settled here, traffic would also be held up by the cattle drives in spring and fall, when teenagers would move herds of heifers up to pastures in Ashby for the summer. But then the highways came, and after that came the computer companies, and after that the housing developments.

First to arrive was a massive Digital Equipment plant that was built at the top of a hill on the eastern side of the stream that defines the eastern edge of the agricultural lands of Scratch Flat. It stood above the surrounding orchards like a cathedral for the new religion of technology. I was a newcomer and a nonentity back then, but I did my best to protest some of these changes. As it turned out, however, there was already a group of dedicated environmentalists in the town. Among other good works, they had formed a land trust in 1962 that is still active and has preserved large sections of fields and woodlands. Lord Findlay, who settled in the town in the early 1970s, became an active member and began working to save more land, eventually designing and maintaining trails for the next fifty years.

But in spite of such efforts, little by little the town began to sell off its birthright, tearing down old barns and two seventeenth-century houses and, in one notorious case, uprooting a beautiful peach orchard that was replaced by a gas station. I discovered a rich metaphor in that particular development. The original orchard had been planted in the mid-1800s, at about the same time that the U.S. Army cut down the thriving peach orchards in Apache territory to deprive the tribe of what had been a sustaining resource.

Monsieur Voltaire would have appreciated the metaphor.

In time, the twenty-first century caught up with the town, and now, although there are still quiet tree-lined roads as well as woods, lakes, and streams, the region is traffic-plagued and pressured by development. The new houses in the town are veritable mansions, most with two or three carbon-spewing vehicles parked in their garages, their huge monoculture lawns maintained with herbicides and pesticides and shorn with overly large riding mowers and leaf blowers that foul the air and assault the former silence. The greenhouse-gas contributions of this development are almost incalculable, quite apart from the carbon release created by the construction of houses, driveways, roads, and parking lots. And this is but a small town located in what is hardly the fastest-growing region of the nation.

The final blow—in my view, at least—was the development of a shopping center placed on a hill that was formerly a horse pasture. City folks used to pull over and admire the herds that grazed there. The shopping center was successful, so much so that traffic became even heavier, and the state felt it was necessary to put up traffic lights on the Great Road. When I arrived here, there was only one light in the town center. Now there are four.

But thanks to the local land trust and other environmental groups, there are still rural aspects to the town. It has two lakes, one of which is surrounded on three sides by protected forest and swamplands and is off limits to motorboats and Jet Skis. There are still orchards, also three thriving vegetable farms, two dairy farms, horse pastures, and farmstands. Still, not long ago, farmstands lined the Great Road, one after another, like the fast-food joints of less fortunate communities. Fortunately, some years back the town prohibited drive-through eateries, so there are no fast-food chains; in fact, there are no McDonald's within ten miles. As on Martha's Vineyard, people who did not know the place in the old days think it's a pleasant rural New England town.

In spite of traffic and noise and the drought, that year brought on a peaceful autumn of mists and fruitfulness. The rains began in early September; the lawns greened up; the yellow leaves of the aspens fluttered in the wet winds. Frogs appeared in the rough uncut sections of the garden; and toads, which had seemingly stayed below ground in the few moist spots in the garden, now emerged to consume their usual diet of pest insects. The apples and pears ripened in the local orchards of the Nashobah Valley, and city-bound families made weekend expeditions to pick the fruits of the summer and take their children on hay rides and enjoy newly pressed cider.

By late October, the garden began slowly closing down. Asters and goldenrod and chrysanthemums bloomed; monkshood flowered; milkweed pods split open; the lawns and the flowering mead in front of the house took on a deep green hue that contrasted with the flaring colors of the maples and the hickories and the oaks.

Leaves started to fall and I left them on the ground, creating in the various garden rooms, a patterned, multicolored Persian Qashqai rug. In certain sections of the garden, inky caps and oyster mushrooms and meadow mushrooms appeared. The air was redolent with the musky odor of old leaves, withered nasturtiums, and a few varieties of roses. Plants sprouted and flowered, faded, set seed, and either stayed green for the rest of the growing season or, as with the early spring ephemerals, slipped back to earth to wait for spring.

Flocks of birds began passing through: first the barn swallows, then the Savannah sparrows, then the white-throated sparrows, and, in late October, great rivers of migrating blackbirds as well as myrtle warblers and tree swallows. Starting in September, the quiet migrations of hawk species commenced and carried on into December, when the red-tailed hawks began moving. At night I could hear barred owls calling from the tree-lined banks of Beaver Brook, and on still clear nights when the moon was full, I sat out on a chair on the lawn to watch the migrations through my binoculars.

Even darkness reveals energetic pulses of changing seasons. Many of the smaller passerine species of warblers migrate by night and rest in the day; and if you're patient and watch long enough, you can see the little dark specks of small birds passing in front of the moon, all of them southbound. And this was only a narrow glimpse of what was happening during the single half-hour or so that I spent out there.

Nights are loud in the garden in late summer and early autumn. Until the first hard frosts silence them, the yard is surrounded by the clicks, chirps, buzzes, and whistles of stridulating insects— katydids, bush katydids, field crickets, meadow crickets, and snowy tree crickets, among them—calling from the shelter of shrubs and unmown grasses. But slowly, as the cooler nights advance, one by one, the sounds of the insects quiet down and then fall to silence. Last to go are the snowy tree crickets, whose flutelike chirps slow little by little until they, too, cease altogether. The phenomenon reminds me of Joseph Haydn's *Farewell Symphony*, in which, one by one, the players in the orchestra rise and leave the stage until the last instruments, the violins, fade into silence.

By day in early autumn the garden is alive with butterflies, including the now sadly diminished populations of monarchs, which, in an epic migration, will manage to get all the way to northern Mexico, crossing the Gulf en route—along with another small garden jewel, the ruby-throated hummingbird.

Also en route in that season are the large green darner dragonflies, which dart across the garden lawns on sunny warm days, sometimes as late as early November. Eventually, they, too, migrate to southern states and then work their way back north in a series of generations.

And after that there comes a slow closing down of the garden as the first light frosts strike, killing the sensitive ferns and the nasturtiums and the cosmos and most of the zinnias and dahlias.

That fall, there was a terrible hurricane in southern Florida, but not a record breaker. There were more floods, though, in Bangladesh, strange late heatwaves in Europe, and bizarre early snowstorms in the Rockies that prefigured a heavy snow season. Climatologists were watching the Pacific weather patterns and began predicting a weakening of the cyclical La Niña and the onset of El Niño, which would mean a general warming in the coming summer and winter—not good news for the already overheated planet. Weather watchers worried that the summer of 2023 could see sustained, record-breaking heat waves around the world. Of course, there were fires in the West, cyclones in the South, fluky monsoon patterns, drying riverbeds, flooding riverbeds, and more dire predictions about the Greenland ice sheets and the thinning of ice in the Antarctic.

Two European glaciers melted almost entirely. New accounts were released about the worldwide decline of insect populations, and UN reports continued to record an increase in atmospheric carbon dioxide levels, despite international regulations.

My old friend Dr. Pangloss Rosen had a heart attack that fall and was admitted to the local hospital. I went to visit and found him tired but resigned to his fate. He had been under anesthesia a few days earlier, and I think he was still wandering around in some dream world.

He seemed at first to be his old self, and fairly alert. After a greeting and a complaint about what he called his "imprisonment," he began a rambling dialectic on heartbeats—the "tides" of the pulsing heart and then the tides of the Earth as a heartbeat. I think he was referring to something akin to the survival of the Gaia.

"All is balance," he said, softly. "The moon-drawn tides. All balance. Earth is balance, and if it falls out of balance, as it has now, it will correct itself. When the prey declines, the predator declines. And when the predator declines, the prey rises again. You know all that, don't you?—you with your animal metaphors for everything. The predatory tide of extinction is now upon us, but it will rise again in some newborn wild Earth."

Why he chose to relate all this rather than discuss his health prospects, I don't know. But that was Dr. Pangloss.

"Sounds good," I said. "But how long will it take for this tide to turn?"

"Who knows?" he said. "Who cares? Henry"—he meant Thoreau—"says time is but the river he fishes in."

He drifted off into a semi-sleep at this point, and I thought I should leave. He wasn't making much sense. But then he woke up again, seemingly revived, and told me to stay on. "There's no one to talk to here. My ex-wife was here, but she just caught me up on local gossip. I want to talk."

He said that the one good thing about being imprisoned is that he had time to think (as if this was a change for him; it seemed to me that all he ever did was sit in a chair and think). He said he wanted to get some of his thoughts out.

I told him I'd been thinking too. I always start getting ideas during the changes of the season. I told him a little more about my work-in-progress and how I had been mulling over the interrelationship between gardens and art. In my view, I said, planting grounds with different species of trees and shrubs in an arranged pattern that pleases the eye is creating a work of art.

"How so?" he asked.

"Well, the early landscape painters, Claude Lorrain and the like, looked out over an unhoused section of wild and tried to capture it on canvas. That, in a way, is what a garden designer does."

"Capture, you say. Why capture? Sounds like the hunt."

This questioning was new for him, maybe the effect of the anesthesia. Usually I was the one who asked the questions.

"Well, maybe that's it," I said. "The Paleolithic cave painters created images of their prey. They didn't paint flowers or trees. Only animals. Drawing them was another way of capturing some essence of them. Some archeologists see it as a magical transformation, a way of ensuring a good hunt."

He knew all this already, but he was uncharacteristically slow to respond. Eventually he said, "They painted lions too. Dangerous lions. Cave bears."

"Well, that's part of landscape art, isn't it?"

"How so?"

I told him that my understanding was that part of the point of the early landscape paintings was to evoke a feeling of the sublime. People used to fear mountains, but then the artists went out and painted them and evoked a feeling, a reaction, fear.

"*Terribilità*," he muttered. I could hardly hear him. "The sublime."

None of any of this was news to him. In fact, I was just reiterating some of his previous lectures.

"What's your point, though?" he asked, brightening. "Gardens aren't scary. Quite the opposite, in fact, as you keep telling me. They are sanctuaries. Safe places."

"Yes, so the gardener is the hunter. She, he, whoever, goes out and tries to capture the 'animal,' the place where they feel safe."

"You're always thinking in animal metaphors, aren't you?" he said again.

His question was an indication of intimacy and friendship—proof, I think, that he was actually listening to some of my own ramblings. I believe his own mortality was catching up to him. But I suppose what he said was true. Thinking about it, it struck me that a garden is not so much about the hunt; it is the result of inspiration evoked by certain places apart, where, for whatever reason, Neolithic people felt the presence of a god or a spirit. So they tried to re-create it. The creation of a place—a sense of place. A grounding. Also, a statement, a way of saying, "This is the way the world should be."

I told him all this, but when I glanced over, I saw that he had fallen asleep again.

"I think I should be going," I said quietly.

"No, no, I'm awake. Tell me more about your garden book."

What to say?

So I told him about experiences I had been having over the past couple of years during the growing seasons, how sometimes, at dusk, I walk out around the garden paths and lose myself. I make a turn and come upon a garden room I don't recognize. It's a surprise, as if I've come to some undiscovered place in the world. I forget that I was the one who created it. That, I told him, is the oneness he is talking about, a re-creation and statement of the ancient communion with nature. I rambled on, thinking aloud to myself about the garden as a statement of the way one thinks the world should be. The gardener as the artist, the garden as a creation or map of your own mind, and how planting is a statement of belief in the future and a commitment to life and clean air and all of the things I had been talking about with him during the past year. At that point, I launched into a long description of my thirty-year project of transforming a sterile forest of white pines into a garden of more than 2,000 species.

I noticed that he had fallen asleep again.

"Anyway, I think I should be going," I said at the end of my speech. I got up to leave.

"Where are you going?" he asked. His eyes were still closed.

"Home, I guess," I said.

"Home," he said. This was a statement, not a question.

"Yes, just home."

He grunted softly. He was clearly thinking something about the meaning of *home*, but before he could express it, a happy nurse came waltzing in and asked how "we" were feeling and told him they wanted to do a few more tests.

It was an abrupt ending. I had the sense that he'd been about to launch into a sweeping overview on the meanings of home or perhaps the idea of place as a statement of the individual within a community or something like that. But it was clearly time to get out. I said goodbye and left.

On the marshes of the river just below the hospital, I noticed a huge swirling cloud of migrating tree swallows. It was autumn.

Winter was coming. I sat in the car for a while and then drove up to a nearby community-supported arboretum and took a walk.

Thinking about old Robert Rosen and all of our past discussions and his recent garbled talk of tides and the course of history, I was struck with the notion that he really was a little like Dr. Pangloss. He did, in fact, have a positive attitude toward the human condition and the fate of the Earth; he believed in hope and the endurance of humanity. His point, nested within his convoluted discussions, was that, unlike other animals, we have the ability to learn from the past and the tools to foresee a future and do something to prepare for it.

Wandering around the arboretum and its collection of diverse flowering plants, I thought about my own garden and started making some really big plans. What else could I do? Other than writing, I don't have any job skills, so I might as well plant roses.

Later that day, perhaps because my old friend was reminding me of mortality, I thought about an experience I'd had while walking around the back streets of an old mill town earlier that autumn. Along the way, I'd come across an empty house that clearly had not been lived in for quite a while. The paint was peeling, the plantings in front had gotten out of hand, and a wisteria was twining in through a broken window. I decided to do a little trespassing and see what I might find of interest in the yard.

In back, I came upon an overgrown rose garden. Some of the bushes were still in bloom; but as I drew closer, I was assaulted by the smell of death, mixing with the scent of roses. I presumed at first that some animal had died in the rosebeds but then realized that the smell seemed to be coming from the house.

I wondered if the person who lived there had died and had yet to be discovered. This had happened a few weeks earlier to one of my brother's elderly neighbors, who'd been living alone.

I learned later that this was indeed the fact. The owner of the house with the rose garden had been a reclusive elderly woman who lived alone. Her sometime groundskeeper eventually found her body. I was reminded of William Faulkner's short story "A Rose for Emily," which describes a similar situation; also, a scene in Giuseppe di Lampedusa's novel *The Leopard*, when Prince Fabrizio Corbera di Salina catches the odor of a dead soldier in his lush Sicilian garden.

It was an ironic experience. The last roses of autumn were blooming without the help of their caretaker. I could only hope that whoever bought the house would be a gardener.

Archeologists have found fossil evidence that roses first evolved between 35 and 70 million years ago, which means that they survived at least one or two periods of mass extinctions. One presumes that long after the deluge of climate change and the end of the Anthropocene era, the indifferent roses may still be blooming.

Epilogue

Everything the various climatological reports of 2022 predicted came to pass in the winter and spring of 2023–24. The year 2023 turned out to be the hottest in recorded history.

I went through the usual autumn chores, banking the roses, raking leaves, preparing the flowerbeds for the traditional long winter, and watching for the first frosts.

The news on the energy front was mixed. Economists determined that solar and wind power were less expensive than oil, gas, or coal and were saving the economy billions of dollars. Thirty percent of the world's electricity was generated by renewable resources, and all looked well for the future, with more wind and solar projects waiting to be developed. But there was a snag. Greenhouse-gas emissions continued to rise and broke another record in 2023. The race was on, and there were serious hurdles.

Weather reporters were told to scale back the climate-change updates. Fossil-fuel interests were working hard to prevent the market from shifting to renewable systems. Company-funded think tanks and trade associations, coupled with skillful advertising, were hard at work disputing the benefits of wind and solar power, questioning the science and the financial benefits of renewable resources, and even claiming that fossil-fuel production did not cause climate change.

This, of course, is an old story, as my Dr. Pangloss pointed out by way of his slavery metaphor. Back in the early 1960s, when research indicted tobacco as a cause of heart disease and cancer, the tobacco companies spread misinformation through advertisements and faulty science. A great irony on that front is the fact that the macho cowboy of those advertisements, "the Marlboro Man," died of lung cancer. Such ironies continue. One wonders how many people will have to suffer through asthma, lung disease, and heart problems before successful remedies to air pollution are introduced.

As far as climate change is concerned, the evidence is in our own backyards. As I've said, twenty years ago, on or near September 18, a light frost would hit the sensitive plants of the garden, and then the weather would warm up again. Around Columbus Day, in early October, there would be a genuine killer frost.

Nowadays we usually get a light frost on October 5 and then, sometime in early November, a serious freeze. But in the autumn of 2023, a light frost in early November knocked back the sensitive ferns and killed the nasturtiums, and then Indian summer settled in: the roses bloomed profusely; the flowerbeds stayed green and continued to blossom. The weather felt more like late summer than autumn, and the warm conditions continued. December arrived, and still no killing frost. Then, finally, a few days in, we had a real frost. Overnight the garden changed—but then it warmed up again, and kept warming. We had a green Christmas and a little snow in early January, enough for me to check on the comings and goings of the local deer, foxes, possums, skunks, raccoons, coyotes, and the occasional bear. As far as I could tell from other signs in the nearby woods, the bears had yet to hibernate because of the prolonged warm weather. Winter proceeded onward in its non-winterlike way. By mid-March of 2024, more than two weeks early, spring began in earnest; and by mid-April, the full garden of earthly delights had burst out.

I went out early one morning, sat down in the little clearing beside the greenhouse, and fell into another one of those drowsy daydreams in which time slows to stillness. The magnolias and the early rhododendrons and azaleas were in full bloom, along with the forsythias and daffodils and narcissi and the tulips and hellebores, the hyacinths and periwinkles and the blue scillas, the weeping cherries and the peaches, apple, and pear trees. And sounding out from the shrubs and trees were the Carolina wrens and the cardinals, the tufted titmice, chickadees, chipping sparrows, mourning doves, red-winged blackbirds, grackles, flickers, pine warblers, ruby-crowned kinglets, everyone in full song.

Could it be that hope is a thing with feathers?

This is hardly the best of all possible worlds, but could it be, I wonder, as Voltaire's contemporary Alexander Pope claimed, that hope springs eternal?

Acknowledgments

In no particular order, I would like to thank the following for their help with this book:

First, the various reporters for *Sanctuary* magazine, for their contributions to climate- related stories between 1988 and 2016.

Also the following, for their conversations dealing with climate changes: Mark Kelly, Joy Reo, Susan Edwards, Robert Rosenthal, Stephen Hecht, Timothy Ahern, James Geraghty, Richard Forman, Daniel Goodenough, Hugh Mitchell, and Chris Hohenemser. And for unofficial climate-related interviews, the random and various street preachers, solar engineers, energy specialists, climate deniers, religionists, hedonists, optimists, pessimists, and futurists who I have talked with over the last twenty-five years and whose names I have forgotten or never knew.

For garden discussions and plant contributions, thanks to Jeannie Abbott, Susan Schnare, Suzan Osborn, Deborah Costine, Beth Van Gelder, Abigail Higgins, Susan Harvey, Lawrence and Kim Buell, Rick Findlay, Andrew Bowers, Leonard Gerwick, David Palmucci, Lelia Leary, Margie Wheeler, Dale Szczeblowski and Laura Howick, and Beverly Bringle, as well as talks with members of the various local garden clubs who have visited the garden over the years. Also, my wife, Jill Brown, for our seemingly endless talks on garden design, flower colors, plant arrangements, and her tireless weeding.

And finally, for our Jack Russell Terrier for his willingness to dig on command perfect holes for tulips and daffodils.